# WEEK LOAN

2003.

# Wireless Personal and Local Area Networks

# Wireless Personal and Local Area Networks

## Axel Sikora

*Department for Information Technology,*
*University of Cooperative Education,*
*Lörrach, Germany*
*&*
*Steinbeis-Technology Transfer Centre for*
*Embedded Design and Networking,*
*University of Cooperative Education,*
*Lörrach, Germany*

**WILEY**

### Other Wiley Editorial Offices

John Wiley & Sons Inc., 111 River Street, Hoboken, NJ 07030, USA

Jossey-Bass, 989 Market Street, San Francisco, CA 94103-1741, USA

Wiley-VCH Verlag GmbH, Boschstr. 12, D-69469 Weinheim, Germany

John Wiley & Sons Australia Ltd, 33 Park Road, Milton, Queensland 4064, Australia

John Wiley & Sons (Asia) Pte Ltd, 2 Clementi Loop #02-01, Jin Xing Distripark,
Singapore 129809

John Wiley & Sons Canada Ltd, 22 Worcester Road, Etobicoke, Ontario, Canada M9W
1L1

Wiley also publishes its books in a variety of electronic formats. Some content that appears
in print may not be available in electronic books.

### British Library Cataloguing in Publication Data

A catalogue record for this book is available from the British Library.
ISBN 0 470 85110 4

Typeset in 10/12 pt Times New Roman.
Translated by TransScriptAlba Ltd., Oban, Scotland, <books@transscript-alba.co.uk>.
Printed and bound in Great Britain by T. J. International Limited, Padstow, Cornwall.
This book is printed on acid-free paper responsibly manufactured from sustainable forestry in which at
least two trees are planted for each one used for paper production.

# Contents

# Preface

Computers and microelectronic devices are increasingly being networked to bring to life *anybody* and *anything, anytime, anywhere* and to add value to devices and applications by enabling them to access information and resources over any distance. *Wireless Local Area Networks* (WLANs) have enjoyed enormous popularity for this purpose, in the last few years, because they provide user-friendly, inexpensive and effective access to applications that are becoming ever more portable or even "pocketable". Since semiconductor technologies started providing the necessary performance it has also been possible to develop highly-interesting wireless networking products commercially.

The subject of WLANs is therefore extremely topical, which poses a fundamental problem both for users of WLAN systems and for the author of this book: the development of very many technologies, products and market participants has not yet been consolidated, and progress and technological innovation in these areas is non-stop. In addition, networking systems companies are involved in intense marketing activities, which makes it even more difficult to understand the situation globally. In the Internet age we are faced with a massive flood of information. In particular, users have to be able to pick out which information is relevant for them, identify false expectations and work out realistic estimates and scenarios based on real-life situations.

The aim of this book is to act as a solid foundation on which you can build to form a comprehensive and up-to-date overview of wireless LAN technology for yourself. For this reason much of the book is taken up by the technological basics, with special emphasis on explaining the differences between the various technologies.

## Structure of the book

As the figure below shows, the book is split into three sections:

- The first chapter deals with the common aspects of wireless LAN networks, and the chapters that follow contain more background information about their structure and physical characteristics.

- Later chapters describe the most important LAN standards that are used in the market today, and will be used in the medium term. These standards are IEEE802.11 (Chapter 4), Bluetooth (Chapter 5), DECT (Chapter 6), HomeRF (Chapter 7) and HiperLAN/2 (Chapter 8).

- In the third, practice-oriented section, Chapter 9 provides an introduction to the structure of WLAN networks, and how to administer them, and describes some sample installations, before selected aspects are discussed in greater detail in Chapter 10. They mainly include safety issues, electro-magnetic tolerance and useful criteria for selecting WLAN products.

*Figure 1:*
*Structure of the*
*book*

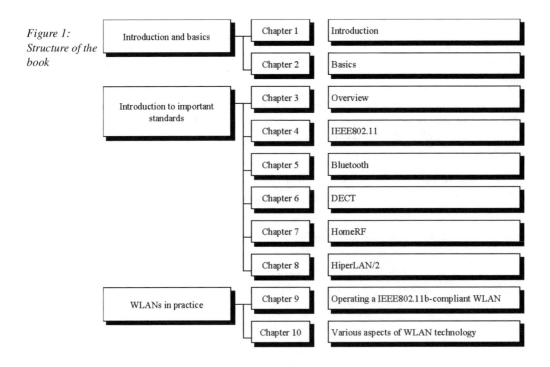

The contents of this book were put together for a series of lectures on "Computer systems" and "Communications networks" at the University of Cooperative Education in Lörrach, Germany (*http://www.ba-loerrach.de*) and for various industry seminars that were held at the local Steinbeis Technology Transfer Centre for Embedded Design and Networking (*http://www.ba-loerrach.de/stz-edn*).

Many enthusiastic and highly-motivated people have helped me to investigate this subject thoroughly and prepare the material and information. In particular I should like to thank Claus Donner and Peter Schwindt from the computing centre at the University of Cooperative Education in Lörrach, Germany, Jörg Luther from the online magazine "tecchannel" (*http://www.tecchannel.de*), Thomas Boele of Cisco Systems GmbH in Hallbergmoos, Germany, Hans-Joachim Drude of Tenovis in Frankfurt, Germany, Peter Schmitz of Network Associates GmbH in Nettetal, Germany, Ralf Keller and Hubert Kraus of Netlight GmbH in Offenburg, Germany and Markus Radimirsch from the Institute of General Information Technology at the University of Hanover, Germany. This edition is based on the German book "Wireless LAN", published by Addison-Wesley Germany in 2001, but with updates in chapters 4 and 10.

Last but not least I must give special thanks to my family once again for their loving support and understanding.

*Heitersheim, Germany, October 2002*


P.S.: The publisher and the author are always pleased to receive comments, feedback and suggestions. Please email them to *asikora@wiley.co.uk*, John Wiley & Sons Ltd.

# 1 Introduction

## 1.1 Definition and restriction

*Wireless Local Area Networks* (WLANs) are standardised network technologies that implement LAN functionality by using wireless data transfer. This primarily includes the broadband connection of microelectronic devices in a range of tens of meters to a hundred metres. (You will find a more detailed definition of local networks in section 3.1.2.)

There are major differences between WLAN technologies and the other technologies called "wireless" technologies. These differences apply even if some of the applications described in section 2.3 can be implemented with those other wireless technologies. Three areas of wireless messaging that differ from WLAN technologies should be mentioned, in particular:

- A wide range of different proprietary systems are available for connecting peripheral devices (for examples, see section 3.1.1). This book does not discuss these systems in any more detail because they do not comply with any particular standard and because they cannot usually be used in a network.

- A number of different radio relay systems are available, that use light waves (laser links) or electro-magnetic waves to link local networks, like in the application described in section 3.1.5. However, these *Point-to-Point* (P2P) or *Point-to-Multipoint* systems (PMP) devices cannot usually be used to create linked networks.

- Digital mobile telephone networks can be used to transfer data, and not just voice traffic. This method is often used for creating a wireless connection to field devices. However, this does not create a local network but a global network with full mobile support. In addition, you cannot generally use this kind of network with a broadband connection. Even the bandwidth of UMTS networks, the technology used for the third generation of mobile telephones, is restricted to a maximum of 2Mbps. In many situations, this bandwidth sinks to values of 384 kbps or even a mere 128 kbps [Sietmann 2001].

## 1.2 Overview of advantages and disadvantages

The wireless transfer of information has potential advantages for all network hierarchy levels. For some time now, personal and local wireless networks have aroused particular interest because the numerous portable and pocketable devices used for voice and data processing or data transfer via wireless transmission media can connect to each other and communicate directly without the need for a cable. This means that *Wireless Local Area Networks* (WLANs) can replace the corresponding cabled solutions on the same network level, in such situations.

As a result the tangle of cables under your desk can be reduced or done away with completely. The extra bonus is that there are no more problems with defective cables.

Problems with incompatible connectors (plugs) or missing cables can become a thing of the past.

With regard to cable connections in the office, if wireless technology is used to provide data transfer in the office and at home it can replace expensive cable installations, depending on the network hierarchy level involved.

In ad hoc networking: the functionality available for actively searching for potential communications partners, and for automatically negotiating transfer and application protocols, are of special interest (see section 1.3.10). Many of the most attractive features of wireless technologies can be used with *Intelligent Networks* (INs) and also implemented in wired systems. This is particularly true if these services occur in the upper protocol layers, regardless of the actual physical transfer route. However, as these services are designed and implemented by the manufacturers of wireless communications systems (and, for the first time in computing, maybe therefore be implemented logically and consistently!), these services are identified with those manufacturer-specific systems.

Using the features of wireless transfer, devices can also be used "on the go". In this case the size of the radio cells, and connections to other systems, depend heavily on which system you use. There are also considerable differences in the speed at which the user can change their location (move about). In addition, as a result of the progressive scaling of semiconductor devices, these products have become increasingly affordable. However, these developments require a high take-up rate and therefore even manufacturers are trying to force the pace of market launch to solve the "chicken or the egg" problem of high numbers of units produced versus low unit costs.

Despite this, the implementation of wireless networks is also subject to a number of limitations:

- *Cost per bandwidth:* Although the cost of wireless systems is going down, the bandwidth for wireless systems is still significantly more expensive than for comparable wired systems. In the case of active components you must reckon on them being a degree of magnitude (10 times) more expensive. However, if you include installation costs, the figures are once again clearly in favour of wireless networks.

- *Availability of bandwidth:* In many cases, not only is the bandwidth for wireless systems more expensive than for wired systems, but less bandwidth is available for wireless systems. In the medium term you can also expect to see costs drop to 10% of their current levels here.

- *Range:* The range of wireless systems is quite restricted in many cases so that the functionality expected by users either cannot be achieved, or can only be achieved to a limited extent (see section 2.4).

- *Electro-magnetic radiation:* Even if the field strengths used for wireless local networks are only 5% of the strength of mobile GSM telephones they increase the electro-magnetic burden on the environment (see section 10.4).

- *Investment security:* At present a wide range of solutions are still available in the market. However, only a few of them will enjoy medium-term commercial success. If investment security is a primary concern, this situation can often delay decisions about future investments.

- *Interference:* Wireless LAN systems transfer data via the air. Unlike wired systems, they use a common medium, known as a *shared medium*, in the sense that not only the stations in one channel, but also different channels and technologies access the same, shared medium. They do not support switching.

- *Ruling bodies and international aspects:* The fact that transfers in wireless systems take place via the airwaves (a public medium) also means there are numerous organisations that set the regulations governing who can use the various frequency ranges. These regulations cover not only the ranges of useable frequencies but also the permitted modulation procedures, maximum transmitter power, compatibility with other systems and many other issues. Even if the average user is not usually interested in the technical details, they must sometimes take into account the fact that they will not be able (or allowed) to use their own system when they travel abroad.

- *Security:* As the range of a radio transmission is not restricted solely to the area in which the potential recipients are located, but is also broadcast to other areas, there is the risk of it being "overheard" by unauthorised recipients. In addition to this, many systems support automated and dynamic logon for the mobile stations. To close these potential gaps in security, existing systems offer a range of different protective measures. However, the actual levels of security these provide can vary considerably.

As always, the fundamental question for a user is how much value they can add to their application by using a wireless transfer technology.

# 1.3    Applications overview

## 1.3.1    Adding value by using a WLAN

Below you will find a number of typical examples to show the value that can be added to actual applications by implementing a WLAN. Section 9.6. describes a selection of real-life installations.

## 1.3.2    Mobile workstations

Providing support for mobile workstations is probably the first use for WLANs that springs to mind. However, this not only applies to the office environment, in which laptop-based applications can easily be supported in the various conference rooms and offices, but also to teleworking (home office) situations in which you could move your workstation to the garden on sunny days.

### 1.3.3    Reducing the number of network cable connections

In wireless networks you will require very few network cable connections for communications (or perhaps even none at all). As a result you can either cut the costs of network cable installations, or avoid them entirely. Although modern office buildings are fitted with conduits that make network cables easy and cheap to install, there are still many, many situations where this is not yet the case. For instance, installing network cables in historic buildings can often be either very time-consuming and expensive or even impossible (for example, if the buildings have protected status such as the "listed" status used in the UK). When new companies are founded, their structure, and therefore their infrastructure, can sometimes change radically in a short space of time. It therefore makes sense to find a flexible solution that suits their needs. Also, in the home, cable channels and empty pipes are only rarely available so it is only possible to install additional network cables either by spending a lot of money, or by having cables exposed to view.

The great benefit in reducing the number of network cable connections becomes especially apparent when a large number of remote devices are to be fitted with network connections. The use of wireless networks can provide a crucial advantage in the context of improving the Internet capability of what are known as "embedded systems" [Sikora 2000 (4)].

### 1.3.4    Training centres

Wireless networks can be implemented in training centres in two ways:

- Nowadays, an increasing number of secondary and tertiary students have their own laptops. This is leading to the demise of stationary computer rooms in training centres. Computers owned by students can be networked quite easily in any training room that is fitted with an network access point (AP).
- The use of wireless networks makes it easier to set up mobile training rooms for special events, seminars or trade fairs.

### 1.3.5    Mobile data entry systems

In many situations data entry takes place on site. Typical examples of this are:

- stock-taking in commercial companies or warehouses
- the analysis of technical devices such as cars or aeroplanes or
- data entry in hospitals, during the doctors' rounds, or in operating theatres.

A direct connection to the corporate network enables these applications to compare and synchronise their data with the central database.

### 1.3.6    Hot spots

Increasingly, waiting times at airports or railway stations or stops at public places are being used for communication via e-mail, or for obtaining information via modern online media. In such situations, wireless networks can provide user-friendly and flexible access to the Internet. It is particularly difficult to implement commercially-viable solutions using, for example, third-generation mobile telephone networks, in small areas that experience high volumes of data traffic (see section 2.1). In these situations WLAN systems are an interesting alternative.

### 1.3.7    External networking

It is extremely difficult to link local networks in two separate buildings in a LAN-LAN connection if those buildings are not sited right next to each other. In this case wireless bridges provide a simple and cost-effective alternative to using the public telecommunications network.

### 1.3.8    Wireless Local Loop

In the same way, WLAN technologies can also be used to supply a wireless domestic connection that runs independently of the public telecommunications network. This *Wireless Local Loop* (WLL), which is used to bridge the "last mile", can be used for both voice telephony and data networks.

### 1.3.9    Industrial applications

Many industrial applications in the manufacturing sector run in an environment that is not suitable for cabling. Frequently it is not possible to use cables due to the risk of damage from heat, or physical damage. In addition, buildings are often so large, or otherwise unsuitable, that it is impossible to install cables cost-effectively. In such situations wireless networks have significant advantages.

Flexible production lines provide another interesting area of application because the structure and configuration of the products that are being manufactured change so frequently. The actual implementation of network connections is itself often extremely time-consuming.

### 1.3.10   Ad hoc networking

Wireless networks can offer the facilities of ad hoc networking in two particular ways, which are described below:

- The original meaning of "wireless network" is based on the development of the Aloha network in the late 1960s (see section 2.1.4). This states that stations with a physical connection automatically recognise each other and can therefore communicate with each other [Frodigh 2000]. In this case, the prerequisite is that the stations exchange information before a network connection is established. In addition, these stations must also have the appropriate, reliable security measures to prevent access by unauthorised stations.

- The concept of ad hoc networking has grown from this initial point. It now extends to cover not only network connections themselves but also applications which can communicate independently with each other. The original meaning was restricted to the lower layers of the network model but the extended meaning now includes all the layers in a network (see section 2.2.1).

## 1.4    Market events

### 1.4.1    Initial situation

Currently, the majority of solutions have only enjoyed limited market success because their active components cost more than the ones used in wired systems. The high number of competing wireless solutions is in striking contrast to the dominance of Ethernet standards in the wired LANs sector. This situation is based on two facts as discussed in the next two sections.

### 1.4.2    The "new market"

Firstly, the wireless market is still in its infancy, although immense growth rates have been forecast for the next few years. As a result many manufacturers are implementing special features in an attempt to get themselves into an advantageous starting position.

However, one characteristic of the wireless market is that almost all manufacturers are aware that a proprietary product will have no chance of succeeding. Not only would users be unwilling to accept it, because interoperability with other devices is a basic requirement, but also the development and marketing costs would have to be so high that the manufacturer could not expect to see a worthwhile return on their investment. For these reasons wireless LANs, in particular, have seen the rise of consortiums that, acting on the basis of "non-profitmaking organisations", aim to help each other by co-ordinating product developments and carrying out industry-wide marketing.

Another specific feature of this market is that almost all the major semiconductor, system and software manufacturers have joined more than one consortium. There are four reasons for this:

- Firstly, the large-scale manufacturers have become so diversified that their different company divisions get involved in different consortiums.
- Secondly, the wireless market is still so fragmented that it is very difficult to forecast the success of a specific technology in the market. For this reason, many manufacturers think it is a good idea to keep more than one option open.
- Thirdly, the applications using the various protocols may differ, resulting in a mismatch of target markets.

- Finally, working in a governing body itself means direct access to current information and topics under discussion, even if there is no intention to exert active influence.

Although the increase in the number of multi-company governing bodies and co-operation does not mean that there are no proprietary products, they are mostly company-specific extensions to existing systems. These implementations are also the result of four factors:

- The market changes so quickly that manufacturers are often forced either to decide to launch their products as soon as possible, even if they do not entirely comply with the relevant standard, so as to ensure the manufacturers retain their market share, or else to decide to stop product development until the standards have been finalised.

- As wireless products currently place very high demands on circuit and system design, and production technologies, some more powerful systems, especially those of early generations, are based on company-specific extensions.

- With an early product launch a company can hope that their devices' lack of interoperability will mean future follow-on revenues when those devices are installed.

- In addition, and this is maybe the central issue, manufacturers of communications products that comply with a standard have almost no way of differentiating themselves from other manufacturers in terms of the actual transfer functionality. The only way the products can provide a significant increase in value is through their additional services. In this context, network administration plays a central role. However, an important point to note here is that it is these very services that can significantly restrict interoperability.

Despite this it has been proven that an effective standard, such as IEEE802.11b for data rates of 11 Mbps, which was introduced around three years ago, is capable of driving suppliers of non-standard products out of the market. Two very different cases can be used here as examples:

- "Radiolan" entered the competition as an early supplier of powerful but proprietary systems. How much benefit the investors who have acquired the rights to the products, and, especially, the brand names, will actually gain in the future is questionable indeed.

- "Proxim" was also a very early player in the market with its proprietary products, but after a long delay it did provide a migration path to products that conform to a generally-accepted standard. This applies both to the company's transmission systems, based on the IEEE802.11 standard, which achieved high data rates of 11 Mbps, and to the migration of its OpenAir-compliant devices, which are now being modified to comply with HomeRF.

As you can clearly see, while all new proposals are based on open standards, proprietary systems are on the retreat.

### 1.4.3   Technologies in competition

The fact that wireless products continue to make the highest demands on the capability and efficiency of hardware and software has a second important effect. At present, the less expensive systems for the mass market in particular are only a compromise between functionality and price. With regard to the demands placed on hardware, the major factor is monolithic integration of high-frequency assemblies, which can offer cost advantages under certain conditions [Sikora 2000 (5)]. Two factors have a decisive influence on product properties:

- The positioning of a system in the network hierarchy, which has an enormous effect on the required performance.
- The time at which the product was to be launched in the market also directly affects the target performance. Bandwidth in particular is directly dependent on the capabilities of the semiconductor technology available at the time.

However there are so many different technologies that this relatively young market runs the risk of frittering away its energy, especially in marketing.

### 1.4.4   Convergence

In the communications technologies sector technologies that were originally independent are increasingly converging with each other. Three examples illustrate that this is not only the case for wired communication protocols:

- *Voice/data convergence:* Some developments are specifically designed for the different traffic requirements involved in voice and data transfer. Examples of this are Bluetooth and HiperLAN/2: Other developments use the protocols defined for voice transfer, such as DECT, for data transmission as well.
- *LAN/WAN convergence:* The technologies used in local and long-distance traffic are also increasingly converging. Currently this is particularly true of technologies that can be used both within LANs and for connecting LAN.
- *Wireless/cordless convergence:* On the other hand many developments in the LAN sector can also be used in global networks. An example of this is the GPRS extensions to the GSM system. Interconnection of services also appears to be something of great interest. HiperLAN/2 can for example be used as an alternative access network for UMTS networks in *hot spots*.

### 1.4.5   Interconnection instead of intermingling

It is absolutely certain that we will see the interconnection of the various networks in addition to the signs of convergence described above. This affects not only the interconnection of wireless and wired networks but also the linking of wireless networks at different network levels, as the following examples show:

- Interconnection of wireless and wired networks.

  This interconnection must take place so that wireless devices can be integrated with those that are more commonly available.

- Interconnection of wireless networks.
  - The integration of DECT and GSM end devices for voice communications seems to be a really good idea. The first products are already available in several regions.
  - HiperLAN/2 has been designed for use as a locally-available access network to the UMTS mobile telephony network.
  - The integration of IEEE802.15-compliant PAN systems in IEEE802.11 WLAN company networks is one of the specific aims of the standardisation process.

Before this interconnection can take place the following prerequisites must be fulfilled:

- *Interoperability:* The systems must have similar network and service layers. Similar service access points also improve the chances of creating a less complicated, modular structure. The interconnection of Bluetooth PAN systems and 802.11 LAN networks could be a barrier to achieving this.
- *End-to-end services:* End-to-end services must be provided. This is particularly true for identification, authentication and routing.

However, in this discussion the positioning of the various technologies and their success in the market are of fundamental importance. Here it should be emphasised that positioning is only partly dependent on current technical specifications. To a great extent this also depends on the additional services provided by system suppliers and on the expansion of these services to comply with future standards. The integration of roaming functionality in the HomeRF 2.0 standard opens up integration options even in larger company networks.

## 1.4.6   Supplier positioning

Despite the relative youth of the market there are clear differences between the groups of manufacturers and other suppliers, and their market positioning, an indication (in this respect at least) of the market's maturity. The following three groups exist:

- The first group of suppliers is involved in developing and usually also manufacturing the chipsets required to achieve the required functionality. This includes the provision both of hardware, in the form of the various silicon chips, and also some of the software that runs close to hardware level. This group can be classified as "chipset manufacturers".
- A second group of suppliers uses these basic elements to make finished, ready-to-use devices and systems. They differ from the first group in that some system suppliers use their own chipsets. Although some of these system suppliers only, exclusively, use their own chipsets in their own devices, others, besides using their own chipsets in their devices, also sell their chipsets to other system manufacturers. These "integrated manufacturers" (of both chipsets and systems) are the most common type of network manufacturers, as network   devices   chiefly   provide   communications   functionality.

Correspondingly, many of these suppliers are involved in network systems that comply with IEEE802.11 or HomeRF , whereas the manufacturers of Bluetooth modules, in particular  are targeting the upgrade market in which devices provide additional functionality. Also, some chipset manufacturers only sell their own chipsets and therefore are not involved in the system supplier sector. These manufacturers are particularly common in those areas in which communications is merely an additional function for devices with a different main functionality. An example of this is the many potential devices, from a wide variety of application areas, in which a Bluetooth connection can be integrated.

- The third group involves regional sales and service organisations that carry out on-site implementation and support.

### 1.4.7    Differentiating between the manufacturers

**Functional differentiation**

Devices that are to function in accordance with a fixed standard offer manufacturers very limited ways to differentiate themselves from their competitors, because they cannot use the device's functionality to do so. Their only means of standing out from the crowd is by providing additional services which extend beyond the basic functionality or the services set out in the standard. In particular the following issues need to be mentioned:

- *Security functions:* For many of the most important technologies the architecture and management of security services are only partly described in the standard. As a result, in this sector we see a range of proprietary solutions created out of necessity that provide very different levels of security.
- *Network administration:* Manufacturers who provide centralised device or user administration usually only cover their own devices and either cannot manage devices supplied by other manufacturers at all, or can only do so to a limited extent.
- *Installation technology:* To simplify the installation of access points some manufacturers integrate the power supply cable in the network cable between the hub and the access point. This means you do not need to call an electrician to install the wireless network. The advantage for manufacturers is that the hub must then support the power connection. This point will become a less important product differentiator as the IEEE802.3af (Power-Over-Ethernet) standard becomes adopted and implemented in future products.
- *Data transfer performance:* However, there is still one more important limitation on wireless networks. As the achievable transfer rate in a number of standards depends not only on the channel quality but also on the quality of the receiver, the systems supplied by different manufacturers can have very different throughput rates. For this reason, the ability of wireless LAN systems to transfer voice or data traffic is also a way in which manufacturers can differentiate their products. This phenomenon is almost impossible to spot

in the case of the active components on wired networks: a 100 Mbit Ethernet network adapter sends or receives a bit every 10 ns, as long as a standard-compliant cable that is shorter than 100 m is connected to it. In the case of wired systems, the standard precisely specifies the properties of the transmitter and receiver, in terms of quality, but in the case of the more sensitive wireless systems it is left to the efforts of the manufacturers to develop receivers that offer the best performance possible.

These, and various other "small print" aspects have an enormous influence on system interoperability. As a result, almost all available test results recommend you only use devices supplied by one manufacturer, despite the fact that all devices are meant to be based on the same standards.

### Product bundling

Many devices offer another value-added service: the option of connecting them to a network. This only partially applies to PCs, which are increasingly being integrated in a network. (To avoid misunderstandings, this book means the following, when it mentions a "PC" (personal computer): it not only means "IBM-compatible" PCs, based on the Intel processor architecture, but also the MAC computers made by Apple. MACs offer some interesting options for integrated WLAN solutions). In contrast, this statement remains true for many other embedded systems. However, it is clear that the aim is to achieve as many connections as possible, to the widest variety of devices, depending on the positioning and functionality of the wireless standard involved.

- Even in the area of network connections the more powerful, and therefore also more expensive, technologies still dominate the cheaper and less complex systems such as *Personal Digital Assistants* (PDAs).

- The simpler, less expensive systems attempt to include lots of new applications that can also be connected to the slightly more expensive devices that are in general use. Examples include operator devices, wireless headsets or household devices.

- Bundling with other network devices can also generate additional revenue. For example, there are now ISDN routers for wireless connection, especially in the smaller office and home office (SoHo) applications market, that would meet the bandwidth requirements. Some products even include the integration of a simple firewall.

### Marketing aspects

The success of a particular product in the market is decided by a multitude of other marketing and sales factors, besides its technical functionality. The following are of particular importance:

- *Positioning:* The clear positioning of different devices in the market helps customers select the best device for their purpose.

- *Portfolio:* On-going support for the entire product range of both active and passive components by the same company makes it harder for other companies to enter a market niche that is already occupied. In wireless LAN systems the types of antenna on offer plays an important role.

- *Bundling:* The bundling of offers together with service providers can make it easier for new customers to enter the market. For example, some *Internet Service Providers* (ISPs) have already launched subsidised devices in the market.

### 1.4.8    Growing market

Despite the difficulties listed above, the market for wireless local networks is forecast to enjoy substantial growth (see [Frost & Sullivan 2001], which broadly agrees with many other forecasts for the chip market for short-range radio). Starting from the already significant revenues achieved by Silizium in 2000 (US$ 2.34 billion world-wide) the market volume is set to increase three and a half times in the next six years (see Figure 2.1). Taking into account the drop in price, even greater increases in unit price can be expected. However, we must not forget that this only concerns revenues at chipset level. The revenues for system manufacturers and service providers will increase in proportion to the level of added value that they offer.

*Figure 1.1: Growth forecasts for the WLAN market [Frost & Sullivan 2001]*

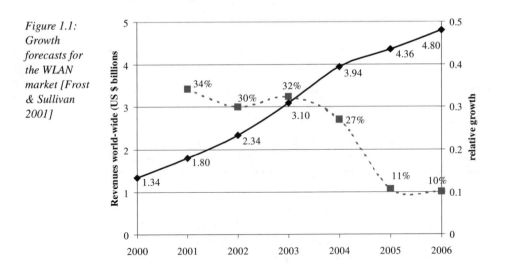

# 1.5    Organisations and governing bodies

A large number of international and national, state and private organisations and governing bodies were involved in setting up the background environment in which wireless transfer systems are used. They can be divided into three categories:

- regulatory authorities for telecommunications which define the framework conditions for the use of the public resource that is the "airwaves".

- standardisation bodies which work out the specifications for the interoperable use of the various applications.
- other governing bodies which (on their own initiative) set out specifications for manufacturers or users.

### 1.5.1 Telecommunications regulators

The telecommunications regulators are basically responsible for allocating frequency ranges and defining the corresponding regulations for their use.

In the USA this is carried out by the *Federal Communications Commission* (FCC) which is the regulatory agency and is directly responsible to the American Congress.

In Europe the frequency ranges are usually allocated by the *European Conference for Posts and Telecommunications* (CEPT) together with the *European Telecommunications Standards Institute* (ETSI). However, it must be remembered that national laws are needed to implement the various recommendations. Once again this leads to incompatibility because of the numerous different historical procedures for creating these laws. In particular, the frequency range of the 2.4 GHz ISM band was restricted for many years in France and Spain because it was reserved for military use. However since the beginning of 2000 the entire spectrum from 2400 MHz to 2483 MHz has now also been made available in France and Spain. In the United Kingdom the national guidelines are published by the Office of Telecommunications (Oftel) which answers to the Department of Trade and Industry in the British government.

For the equally vital Japanese market the *Radio Equipment Inspection and Certification Institute* (MKK) acts as a regulator.

As a result the regulators can sometimes exercise enormous influence on the competition situation because of the importance of their job. A clear example of this is the supplement to the regulations governing the use of the 2.4 GHz band for the Frequency Hopping Spread Spectrum procedure (FHSS) issued by the FCC in August 2000. As a result of this supplement, HomeRF systems from version 2.0 onward can achieve data rates of 10 Mbps (FCC Part 15).

### 1.5.2 Standards and standardisation committees

According to the definition created by the *International Organization for Standardization* (ISO), which was founded in 1946 as part of UNESCO, and since then has acted under the supervision of the United Nations, a standard is a

> *technical specification or other publicly-available document that has been produced in co-operation and general agreement with all affected interest groups, based on consolidated scientific, technological and experimental results, designed for optimum common use and has been accepted by a governing body that is recognised at national, regional or international level* (quoted from [Walke 2000]).

Three central points must be emphasised here:

- the inclusion of affected interest groups
- publication by a recognised standardisation committee and
- public availability

The so-called "company standards" either fail to meet these three requirements or only do so to a limited extent. For this reason, they are not really standards according to the official definition, but instead they are only *de facto* standards.

You will find an overview of the standardisation committees that are active in the area of telecommunications (and the numerous interconnections between them) in [Walke 2000]. The following standardisation committees are of particular importance:

- The *International Telecommunication Union* (ITU), which was founded in 1865 in Geneva and is therefore one of the oldest international governing bodies. The ITU radio communications section (ITU-R) began its work in March 1993.
- The European Telecommunications Standards Institute (ETSI), *http:// www.etsi.org*, a non-profit organisation whose aim is to define long-term standards in the area of telecommunications for Europe and beyond. ETSI has almost 800 members from 52 European and non-European countries. They include manufacturers, network operators, users and national committees. For example, the interests of the United Kingdom are represented by the Department for Trade and Industry.
- The *Standards Association* (SA), *http://standards.ieee.org*, of the world's largest professional association for electrical technology and electronics, the *Institute of Electrical and Electronic Engineers*, Inc. (IEEE). It unites international members with the goal of agreeing standards for an endless variety of technical areas and launching them on the market. The IEEE is especially well-known in the networking sector in which its IEEE802 standard describes local networks, and its IEEE1394 standard describes a fast Firewire bus for communications between multimedia devices.

### 1.5.3   De facto standards and interest groups

A company or company association may develop their own de facto or company standards for a number of reasons. The following can be seen as the greatest disadvantages of the official standardisation committees:

- they take a long time to define and release standards
- they are subject to very many influencing factors, which may contradict their own interests
- they have a responsibility for transparency.

However, the last point is pretty much meaningless. In recent years (and not only in the communications technology sector) more and more organisations have started to believe that only open standards, supported by as many influential and powerful companies as possible, have a chance of success. As a result, these company standards will either be accepted by the standardisation committees or will provide serious competition for them for some time to come.

Examples of these company associations, usually organised as *non-profit organisations* (NPOs) include:

- *Bluetooth Special Interest Group* (BSIG)
- *Home Radio Frequency Working Group* (HomeRF WG)
- *DECT Multimedia Consortium* (DECT-MMC)
- *HiperLAN/2 Global Forum* (H2GF)
- *Wireless LAN Association* (WLANA) and *Wireless Ethernet Compatibility Alliance* (WECA)

# 2 Basics

## 2.1 History of wireless messaging

### 2.1.1 Wireless vs. wired

Usually, but not always, wireless information transfer was preceded by information transfer over wires, as the following two examples show. The spread of wired telephony started at the beginning of the 20th century whereas mobile, wireless telephony has only been able to reach its current popularity due to the introduction of digital technologies at the beginning of the 1990s. Even for local data networks, wired technologies were used first, before the late 1990s brought in suitable wireless technologies for creating wireless LAN systems to provide an effective alternative.

However, all this only takes into account the commercially-implemented systems in each case. Interestingly, wireless systems were important predecessors in both cases, and these systems also anticipated the main properties of wired systems, as the next few sections show.

### 2.1.2 Optical telegraphy

Before wired telegraphy became widely used, from the mid-19th century onwards, optical telegraphy was used in particularly vital areas. This technology was, in its broadest sense, also a wireless transfer system that used electromagnetic waves.

Although the roots of this type of messaging system reach back into antiquity, optical telegraphy systems first became widely used at the start of the 19th century. In the first recorded use of optical telegraphy the news of the fall of Troy in the 12th century BC was transmitted by signal fires over a distance of 500 km from Asia Minor to Argos in the Peloponnese in a single night [Kaiser 1999]. Even if you doubt whether this legend is true, both the Greeks and Romans were familiar with the principle of an optical signal system that used relay stations.

In Britain, at the start of the 19th century, a telegraph system was set up from the South Coast of England to London to provide a means of sending a fast warning to the authorities in case Bonaparte's army crossed the Channel. In France, slightly later, messages were transmitted via a nation-wide network of relay stations with Paris at their centre.

Optical telegraphy has two fundamental advantages: optical signals are easy to generate, and the human eye can see these signals at a great distance. However, they did have some limitations. For example, the transmission channel could sometimes be of very poor quality, especially during the day or in foggy weather.

The original advantages of using human beings to send and receive messages, and the associated low data rate, were both lost with the introduction of wired electrical telegraphy.

In this context there were considerable technical developments in the area of transmission protocols and encoding, early on, which were used later by wired telegraph systems.

### 2.1.3    Using electromagnetic waves

Electro-magnetic waves were discovered in the mid-1880s in the experiments of Heinrich Hertz (1857–1894). After a few diversions Guglielmo M. Marconi (1874–1937) made the first attempts to transmit messages in the mid-1890s. As early as 1896 the first message was transmitted successfully over a distance of 3 km [Kaiser 1999]. However, the first use of telegraphy with electro-magnetic waves was in long-range networks.

### 2.1.4    The wireless precursors to Ethernet

Even in the area of local networks you will find a wireless predecessor for the ubiquitous Ethernet standard based on IEEE802.3. Professor Norman Abramson, who left Stanford University in the late 1960s to join the University of Hawaii, wanted to find a way of connecting to the *Arpanet*. This was an information network which, at that time, only existed on the American mainland, and was an ancestor of the Internet. The only way he could find to connect to it from Hawaii was with a wireless data transfer system. The result was the Aloha network [Abramson 1970], appropriately named after the Hawaiian greeting "Aloha!", meaning (in this context) "Hello and welcome!". The central features of the Aloha network are:

- all stations use a common transfer channel. This feature is called *Multiple Access* (MA).The prerequisite is that the stations are assigned addresses because this type of communication involves more than two participants.

- each station monitors the activity on the transfer channel. If, as a result of this *Carrier Sense* (CS), a channel is recognised as free, the station can start its transfer. The transfer can then start at any time (*Pure Aloha*). However by introducing time slots, which all the stations use, you can significantly reduce the duration of potential collisions and therefore considerably increase the channel's usability. Accordingly, later versions included the development of *Slotted Aloha* which meant that a station could only start its transfer at the beginning of a time slot.

- the fact that stations could start their transfer when the channel became free had the result that two stations could start messaging simultaneously or very nearly simultaneously. The signals overlapped in the common transfer channel and the receivers were no longer able to separate the two signals. In order to identify this type of collision on the transfer channel, and the resulting loss of data, each station was required to confirm that it had successfully received a data packet, by sending an *Acknowledgement* (ACK) signal. The sending station could not delete the sent data packet from its output buffer until it received the acknowledgement that the data had been successfully received by the recipient station. If the sending station did not receive an acknowledgement within a pre-defined time period it sent the packet again. For the waiting time until the start of the second attempt, a comparatively complex backoff algorithm was used to try and reduce the probability that two stations whose packets had already collided at the first attempt would also disrupt each other during subsequent attempts. This was known as *Collision Avoidance* (CA), and some systems still use it.

The properties of the Aloha network, especially those involving the backoff algorithm, were adopted by Dr. Robert Metcalfe of Xerox Palo Alto Parc to use a local network to link several Alto Stations with each other. These were early ancestors of the PC and had a graphical user interface, a mouse and a local printer connection. The first generation of this network, which operated with a transfer rate of 2.94 Mbps, (the processor of an Alto Station was clocked at 2.94 MHz), was also first called an *Alto Aloha network*. Shortly after this the technology was given the made-up name "Ethernet" (for marketing purposes) to make it clear that it was not exclusively for use on Xerox computers. Later on Xerox gave up using their brand name altogether. The IEEE standardised the technology as standard 802.3 [Spurgeon 2000].

During the development of the Ethernet the Aloha network's acknowledgement mechanism was dropped in favour of a *Collision Detection* (CD) mechanism. However, as you will see in section 2.10.2, because this type of collision detection cannot be used in wireless distributed systems, the acknowledgement mechanism principle is now in use once again, even in the most up-to-date wireless LAN systems.

## 2.1.5   Modern use of optical systems

Optical systems are still in use, even today, despite the growing spread of wireless transmission systems based on electro-magnetic waves. Laser links are used to connect physically separate local networks as part of a LAN connection. In the 1990s the IrDA standard was drawn up to govern the connection of devices in a local environment to form personal area networks. This standard became particularly widely used to provide support for infrared connections from mobile telephones and PDAs to laptops.

The latest developments indicate that light waves can be used to provide an easy-to-use and effective point-to-point link between stations if powerful-enough digital signal processing systems are also in place. This is especially true if adaptive procedures such as micro-mechanical mirrors are used. These types of technology could represent an interesting supplement to future wireless LAN access point technology, using the diffuse emission of electro-magnetic waves, because it can support very high data rates [Texas Instruments]. At present micro-mirrors achieve a positioning accuracy of 5 nm, and can then be used to bridge distances of up to 50 m.

## 2.2   Technical communication models and classification

### 2.2.1   Reference models

A complex process is required before you can achieve information exchange between communications partners that achieves maximum success and efficiency and is, as far as possible, automated. To optimise the structure and creation of a modular system this process is divided into functional elements in accordance with a hierarchical layer model. The classic *Open Systems Interconnection* (OSI) model for the connection of open digital systems, which was developed in the 1970s by the *International Organization for Standardization* (ISO), uses seven layers, as shown in Figure 2.1.

*Figure 2.1: ISO/OSI reference model for communication between technical systems*

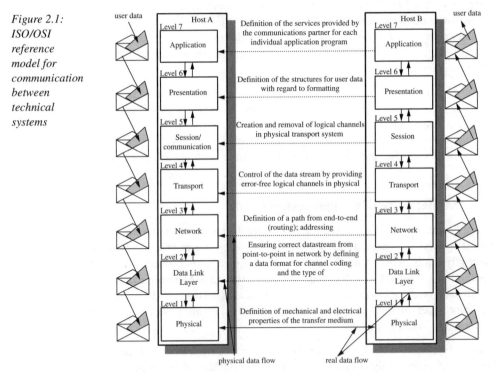

Communications are based on a sender and a receiver, both of which are covered by the term *station*. In this case a station is any technical system that can send or receive messages in accordance with a specific set of rules.

In every station each layer, apart from the highest layer, provides *services* to the layer immediately above it. Before it can carry out these services the layer requires information which is transferred when an instance of that particular layer is created. In many software-based layer implementations you can visualise a layer as a procedure which is called, with a set of parameters being transferred after it is called.

A significant advantage of the hierarchical model is that its functionality is abstract. The calling layer is not at all concerned with the internal implementation of any other layer. Even if, in some situations, the layer has to activate the same mechanism to call other layers in order to provide the service, the calling layer remains unaware of this fact. However, the price to pay for the advantages of this modularization and abstraction is that the actual transfer is less efficient. A crucial point to take into consideration is that each layer involved in communications must add its own frame with control and address information to each data packet transferred from the layer above it (see Figure 2.2). Therefore, if a transmission protocol or a protocol family involves several layers, you get what is known as a *protocol stack* because of the hierarchical structure.

*Figure 2.2: Addition of address or control information by each layer*

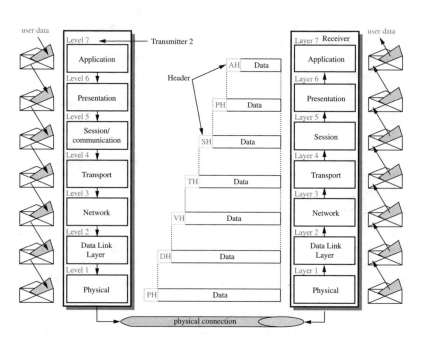

*Each layer adds more address or control information to the data packet.*
*This information is usually in the form of a header and leads to an increase in the volume of data that needs to be transferred.*

In real life this hierarchical structure has the following effects on the use of wireless transmission systems:

- If the service access point for a function remains unchanged a wired transfer route can simply be replaced by a wireless transfer system. In this case the exchange of media is entirely hidden from the higher-level layers. You only need to install a special hardware driver.

- However if the wireless transfer provides a different service access point in which, for example, additional parameters are to be transferred, the layer that calls the service access point must also be modified accordingly.
- If even more additional services (which are not limited to the two lowest layers) are involved with the changed medium, the corresponding applications must also be modified immediately. In most cases it is a good idea to integrate all this with the operating system and create an *Application Program Interface* (API).

In recent years the model has been reduced to four layers as a result of the popularity of the Internet protocol family and the use of the TCP/IP protocol stack. However, people still talk about the seven-layer model.

### 2.2.2    Classification of channel access

In the organisational development of communications, one of the central tasks is to find an answer to the question of who can access the transfer channel and when they can do so. An endless variety of answers have evolved to allow this to be implemented technically, just like the many, many ways that human beings have found to communicate with each other. All these implementations have one thing in common: they try to restrict the number of rules they use to the minimum, and as far as possible attempt to avoid dealing with exceptions. Another aspect to consider is that each time a participating communications device negotiates transfer parameters, this increases traffic on the medium.

Some of the possible implementations have already been described as examples in section 2.1.4.

#### Centralised vs. decentralised distribution

It is essential to ask who actually manages the accesses. Here, there is a choice of centralised or decentral management. In centralised management one central station, the *master*, assigns the channel to other stations. In decentral management, all the stations share responsibility for assigning the channel.

#### Deterministic vs. non-deterministic assignment

In addition it is also important to define how the channel is assigned. In a deterministic assignment process each station has a pre-defined maximum length of time until the next access. In non-deterministic assignment no such maximum length of time is defined. This is not dependent on whether centralised or decentral assignment is used. An example of decentral, deterministic assignment is the IEEE802.5-compliant Token Ring protocol. Examples of decentral, non-deterministic assignment include Ethernet, which complies with IEEE802.3, and wireless LAN, which complies with IEEE802.12.

#### The introduction of time slots

As already described in section 2.1.4, giving the various stations synchronised, timed access is a way of reducing the probability of collisions. This is why almost all modern transmission protocols use what are known as *time slots*, which are a fundamental time unit used in data transfer. This also requires the stations' functionality to be synchronised.

### 2.2.3 Physical communication model

If we look at physical communications in general terms we can identify three separate elements: the sender, the channel and the receiver. The elements of a physical layer shown in Figure 2.3, below, also play a role in the implementation of senders and receivers, as do the logical layers described in section 2.2.1.

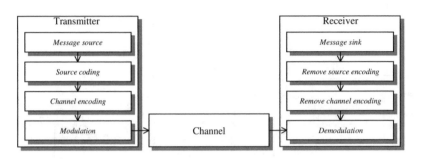

*Figure 2.3: The communication model for technical implementation in the physical layer*

In the source code the source's message flow is represented by the lowest possible rate of digital symbols with the aim of minimising redundancy or irrelevance. This book does not provide a more detailed description of source code.

On the other hand, channel coding has the job of rearranging the code words in the source code to find the best possible match for the channel. Redundancy is usually added deliberately so that the receiver can identify it and also, if possible, correct transfer errors. As the wireless transfer channel is very sensitive to interference (see section 2.4.3), channel coding is used a great deal. Frequently-used procedures include *Forward Error Correction* (FEC) and *Cyclic Redundancy Check* (CRC).

Analogue signals such as voice, music or image signals are usually located in the baseband, i.e. in the low pass range. In contrast, the vast majority of practical transfer channels are band pass channels so the sent signal must be moved to another part of the spectrum. The classic solution to this is the modulation of a sine wave carrier by the source signal. This changes the times of the amplitude, frequency or phase parameters [Kammeyer 1992] (see section 2.5).

### 2.2.4 Communications classification system

Communications via technical networks can be classified according to a range of fundamentally different approaches to implementing them. These are briefly described below because they have a direct effect on the implementation of wireless networks.

### Circuit switching vs. packet switching

In *circuit switching* all network subroutes are physically connected to an end-to-end connection, as shown in the upper half of Figure 2.4. This creates a fixed transfer channel with a dedicated bandwidth. Circuit switching is historically linked to the development of voice telephony where, in earlier systems that were based exclusively on copper wires, a physical connection was switched for every transfer channel.

In the age of digital systems, circuit switching often also takes place via the periodic reservation of time slots, creating a time multiplex, as shown in the lower half of Figure 2.4.

*Figure 2.4: Diagram of circuit switching*

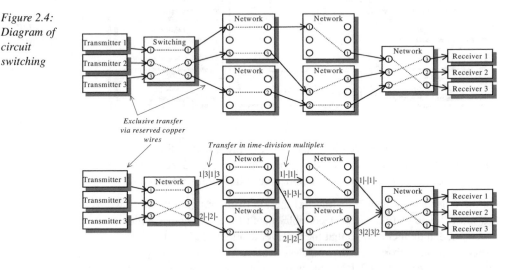

In *packet switching* a packet is not created until there is data to be sent. The packet contains the necessary address and control information and is then sent into the connection network. In this way several stations are able to share the same connection resource. This is shown in the diagram in Figure 2.5.

*Table 2.1: Advantages and disadvantages of circuit switching*

| Advantages | Disadvantages |
|---|---|
| Guaranteed bandwidth. | The circuit is reserved for one connection and is not available for any other services or connections. |
| The devices are easy and inexpensive to implement. | It is only really practical for creating point-to-point connections. |

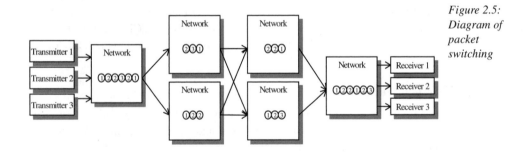

*Figure 2.5: Diagram of packet switching*

| Advantages | Disadvantages |
|---|---|
| Multiple access by several stations to one line. | It is not easy to guarantee quality of service. |
| Traffic streams can be divided up among several connections (load balancing). | Every packet must contain control information. This reduces the usable bandwidth. |

Table 2.2: Advantages and disadvantages of packet switching

At this point it is also worth mentioning another difference between circuit switching and packet switching. At first glance the temporally-multiplexed transfer of different switching systems appears to correspond to packet switching. However, there are two crucial differences:

- In circuit switching the timeslots are reserved periodically. They are then assigned to the corresponding transmission routes, even if no information is to be transferred. In contrast, packet-switching systems only require data when information is to be transferred.

- In packet switching each packet must contain address and control information so that it can reach its destination. The fixed assignment that occurs in channel allocation means that this additional information is no longer required in the packets.

**Connection-oriented vs. connectionless communication**

In connection-oriented communication (*connection-oriented mode*) there are three basic phases: connection set-up, data transfer and connection closure. There is no difference whether the connection is present physically, as part of circuit switching, or logically, in the sense of a message transfer. A typical example that uses a packet-switching network is the *Transport Control Protocol* (TCP) in the Internet protocol stack. This protocol allows data to be transferred completely, in the correct sequence, and without errors, in accordance with a specific connection set-up.

In contrast, connectionless communication (*connectionless mode*) is purely for setting up physical or logical connections. Chunks of data (datagrams) are sent out into the network without any prior communication with the receiver. Connectionless services are usually based on the fact that logical connections are implemented in the higher layers. As a contrast to TCP there is the *User Datagram Protocol* (UDP), a connectionless protocol that is based on the transfer of simple datagrams.

**Timing characteristics**

In the context of communications data networks the term synchronous transfer (from the Greek words *syn*, meaning "with" and *chronos*, meaning "time") means either a datastream that is transferred using a fixed clock rate, or a procedure in which a process with a particular event is integrated into another process, i.e. synchronised.

On the other hand, isochronous transfer (from the Greek *iso*, meaning "same") refers to processes that must run simultaneously to be successful. For example, isochronous transfer is required to transfer speech or moving images.

The third type of transfer, asynchronous transfer (from the Greek *asyn*, meaning "not with"), is quite distinct from the others. Firstly, asynchronous transfer involves the transmission of data independently of other signals, such as a clock signal, and in addition it can also mean the independence of processes.

**Symmetrical vs. asymmetric communications**

Symmetrical communication describes the exchange of data between two identical communications partners. For this to be successful the same transport characteristics are required in both directions. Typical examples are the classic telephone networks, and also connections between two company subsidiaries that are used to synchronise databases.

Asymmetric communication assumes that the two communications partners have different traffic levels. Typical examples of this are client-server services. For example, if a web client (http client, web browser) accesses a web server (http server), a request for the server's service is sent to it using as small a volume of data as possible, and it replies to the request with a larger volume of data.

## 2.3     Demands on transfer networks

### 2.3.1     Traffic types

In addition to the classification system described above the demands on a network also need to be categorised according to the different types of traffic present. In particular, it is important to highlight the difference between voice and data transfer. Here it is clear that wireless networks are expected to do something that wired networks have until now either not been capable of, or only to a limited extent: everyone knows how hard it is to transfer voice data over an Ethernet IP network!

The quality of service at network level is defined by four fundamental parameters:

- Data rate (bandwidth)
- Delay time (latency time)
- Jitter
- Loss rate.

Many protocols have already taken the particular quality requirements of different traffic types into account. For pure data applications (classic data transfer) high data rates are required for short and medium-length transmission times. In the case of propagation delay time, if present, the only important thing is that the entire process must complete within a acceptable period of time. Data losses are in no way acceptable. High bandwidths are usually not required for voice transfer. However, there are a few special requirements concerning the transfer latency time and jitter. These can usually only be met by reserving pre-defined channels. On the other hand the meaning of the voice message is still clear, even if a few bits go missing during the transfer. By contrast multimedia data, as a combination of moving images and sound, presents a different combination of demands for applications such as transferring films. It requires high bandwidths along with low jitter levels. However, the amount of latency time is of secondary importance. To a certain extent the loss rate is not critical, because the human eye can "fill in" missing or incorrect pixels.

## 2.3.2    Transfer speed

The bandwidth of a transfer channel is usually described by the bit rate in bits per second. If, however, several bits are grouped in one character or symbol, as occurs in a multi-value transfer, (see, for example, section 2.5.2), the symbol rate in Bauds is used as well:

1 *Baud* = 1 *character/s*

When describing the performance of transfer networks you should take particular notice of the transfer efficiency as shown by the difference between the gross and net data rate.

A data rate of 10 Mbps simply means that a new bit is output to the channel at an average rate of every 100 ns. However, in real life, this gross data rate is hardly ever achieved. Here are the basic reasons for the reduced net data rates:

- Time slots are not used. It sometimes happens that no station wants to access the channel at various points in time. In many transmission protocols collisions between two stations may occur which then disrupt the data packets and make a new transfer attempt necessary.
- The stations must also deal with protocols as well as the data itself. For example, in packet-switching networks, address and control information is added to the actual data, so that packets can find their way through the network (see Figure 2.5). The stations exchange information about their own

particular configuration. This communication usually takes place via the normal transfer channel (*in-channel administration*). Therefore these times are not available for the actual data traffic.

# 2.4    Properties of electromagnetic waves

## 2.4.1    Maxwellian equations

Unlike in cabled networks, electro-magnetic signals in wireless systems are transferred through the air. The electro-magnetic waves used can be explained by the equations defined by James C. Maxwell (1831–1879) which predicted the emergence of electro-magnetic fields in the presence of accelerated electrical charges. You will find an introduction to Maxwell's equations in Appendix A.1.

## 2.4.2    The diffusion of electromagnetic waves

According to Maxwell's statements, electro-magnetic waves can be generated by electrical energy in cables whose current varies in mnb time. If the length of the cable is of the same approximate magnitude as the wave length then standing waves can be formed. If this occurs, and the power is strong enough, the electro-magnetic waves can separate from the conducting cable.

However, if this happens, extraordinarily complicated descriptions are needed both for the characteristics of the near field in the separation process area and the far field, which describes propagation in space. The electro-magnetic waves are diffused as ground, surface, space or direct waves, according to their frequency and associated wave length, as shown in Figure 2.6. However, using the Maxwellian equations to calculate them is an extremely complicated procedure and is not commonly used if either the geometry or material characteristics are unknown or only partially known. This is why models that only take the basic properties of the radio channels into account are so widely used. In this case the complexity of the models is greatly influenced by the chosen level of abstraction.

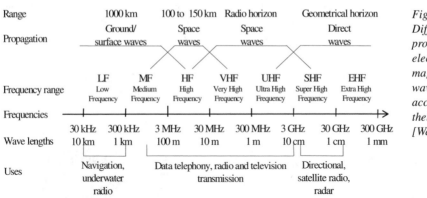

*Figure 2.6: Diffusion properties of electro-magnetic waves according to their frequency [Walke 2000]*

The list that follows contains some more notes to supplement Figure 2.6:

- As a general rule, the higher the frequency, the shorter its range. This is because of the attenuation of electro-magnetic waves, which is described in greater detail in section 2.4.3.

- Low-frequency waves are diffused (spread) as ground or surface waves. This means that they follow the curvature of the earth and can therefore reach great distances.

- Airwaves are mainly generated at higher frequencies. They can reach ranges of from 100 km to 150 km because the waves are bent by the troposphere or reflected by the ionosphere. As the frequency increases this effect diminishes, and therefore the distances the frequency can bridge also reduces.

- Waves larger than 3 GHz spread as direct waves and are therefore limited to the geometric (optical) horizon. They reach the distance known as the *Line Of Sight* (LOS) limit. This is especially true for electro-magnetic waves with frequencies in the terahertz range, which include visible light. Section 2.4.4 describes a few aspects of the diffusion of electro-magnetic waves in the infrared range.

- The distance that can be achieved also depends on the diffused power per unit of volume. The effects that can be achieved by reducing the solid angle are described in greater detail in section 2.9.2.

## 2.4.3   Limitations on the diffusion of electromagnetic waves

The spatial diffusion of electromagnetic waves is limited by a multitude of effects. These include, in particular, damping or attenuation, dispersion, reflection and absorption.

Attenuation reduces the field strength during the mutual generation of the electrical and magnetic curl fields. In this instance the attenuation depends on many different factors:

- Attenuation can only be zero in an ideal vacuum.
- Even very small particles, no matter what form they take, can attenuate electro-magnetic waves. This includes other gas molecules, drops of liquid such as rain or mist, and solid particles or dust.
- Attenuation is strongly influenced by the frequency.
- Larger obstacles can cause massive amounts of attenuation. Table 2.3 shows some examples for the 2.4 GHz range. These are typical values that should only be used for initial estimates.

*Table 2.3: Typical attenuation co-efficients in the 2.4 GHz range*

| Obstacle | Typical attenuation | Source |
|----------|---------------------|--------|
| Windows in brick walls | – 2 dB | [Compaq] |
| Brick wall near a metal door | – 3 dB | [Compaq] |
| Wooden internal door | – 4 dB | [Andren 2001] |
| Partition wall (plasterboard), office wall | – 6 dB | [Andren 2001] |
| Metal doors in an office wall | – 6 dB | [Compaq] |
| Light concrete wall | – 12 dB | [Andren 2001] |
| Metal doors in a brick wall | – 12.4 dB | [Compaq] |

Examples of the complex relationship between attenuation, frequency and atmospheric composition are shown in Figure 2.7, on the next page.

If obstacles are present, dispersion, reflection and absorption lead to a splitting of the original wave. As a result, the majority of signals reach the receiver. However, the differences in runtime and frequency shifts can result in high levels of interference. Figure 2.8 shows an example of multipath propagation in a closed, internal space. This is clearly particularly critical when an antenna with isotropic emission characteristics is used (see section 2.9.1).

It is especially important that these effects are taken into consideration when setting up and operating indoor radio networks as the number of obstacles is especially high in those cases. Just as you need to take into account the physical topology when setting up *outdoor* mobile radio networks, you will also need to work with the statistical description of transfer channels [Walke 2000] and empirical models in particular, even when planning *indoor* radio networks. You will find typical calculation examples, which also illustrate the complexity of making this type of estimate, in [Constantinou 1994]. Basically the reflection properties depend on the material and surface features of the obstacle, on the frequency and polarisation of the wave and on the angle of incidence [Lebherz 1990].

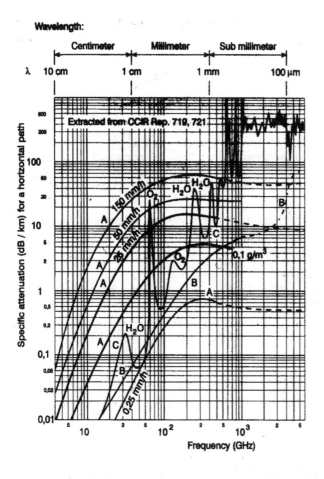

*Figure 2.7: Attenuation of electro-magnetic waves depending on the frequency and atmospheric composition. Taken from CCIR Rep. 719, 721, Figure from [Walke 2000]*

Attenuation due to gaseous constituents and precipiation for transmission through the atmosphere

| | | |
|---|---|---|
| Pressure: | Sea level: 1 atm (1013.6 mbar) | A: Rain |
| Temperature: | 20°C | B: Fog |
| Water vapor: | 7.5 g / m$^3$ | C: Gaseous |

However, as a precise description of the environment involves a great deal of time and effort, generally-applicable models are used to qualify the systems. Three models are most commonly used [Halford 2000]:

- A 2-ray model uses a direct beam and a reflected beam of the same intensity. In this case the most important parameters for the reflected beam are the average and maximum propagation delay time and the standard variation of the propagation delay time.

*Figure 2.8:*
*A number of*
*propagation*
*paths in a*
*closed, internal*
*space*

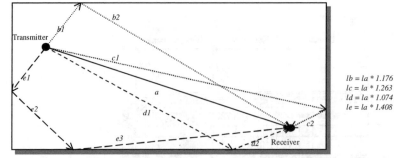

*As a result of the different lengths of the propagation paths the signals reach the receiver at different times. The relative lengths in comparison to linear propagation are shown in the overview.*

- The IEEE802.11 specification recommends an exponential model that is especially suitable for software modelling. In this case the runtime paths are distributed evenly and amplitude falls exponentially.

- As a compromise between the rather simplistic 2-ray model and the exponential model, which is too complex for many applications, the three scenarios defined in the *Joint Standard Committee* model (JTC) are used in real life.

In this context it is useful to define the term "radio cell" (often abbreviated to "cell"). A radio cell represents "the spatial area within which stations can communicate". However, because transmissions in one radio cell can take place over several physical channels, we also need to introduce the term "logical cell", also known as a "channel cell". This refers to "the spatial area within which stations can communicate via one transfer channel".

## 2.4.4    Using light for communications

As already mentioned in section 2.4.2, light, which is an electro-magnetic wave in the terahertz range, is largely non-directional. This is particularly true of the laser light sources used in the communications industry. This results in several various advantages and disadvantages:

- For a point-to-point connection between two stations or networks with a line of sight connection, the low dispersion of light is a very positive factor, as high blanket coverage can be achieved at the transmitter even if its transmitter power is relatively modest.

- Laser diodes of the type used for connections over short distances can be manufactured very cheaply.

- Although the emitter can use a number of methods to increase the beam width, in almost every case the stations must be aligned with each other. One of the main reasons for this is that the beams are damped (attenuated) too much after they have been reflected around most obstacles, and this prevents the receiver from evaluating them.

- The only practical way to create connections between several stations in flexible networks is to use a combination of point-to-point connections.

As a result, it has been possible to use infrared interfaces to link two devices at various speeds for a specific length of time and for a range of applications, as described in the IrDA standard. However these devices are limited by their lack of networkability, and difficulty to align, making them unsuitable for many applications. The advance of wireless networks, as presented in this book, appears to have brought the expansion of the use of infrared-supported links to a halt. For this reason infrared technology and standards are not discussed any further in this book.

Some applications that use light are described in section 2.2.4.

## 2.5    Digital modulation technology

### 2.5.1    Basic principles

Digital signals cannot be transferred via a common air interface in the baseband frequency. There are two basic reasons for this:

- The frequency spectrum of this type of digital signal is very wide, usually even unlimited. If you use a Fourier analysis to calculate the frequency spectrum of a periodic square-wave signal you get the result shown in Figure 2.9. As a result it would not be possible to restrict a transmitter to a fixed frequency range, to allow different radio systems to work alongside each other without errors.

- The transfer properties of the air channel also depend greatly on the particular frequency, as described in section 2.4.2. For this reason the receiver will only receive a very distorted signal.

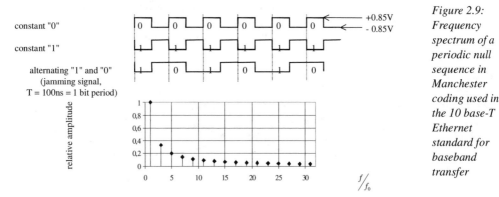

*Figure 2.9: Frequency spectrum of a periodic null sequence in Manchester coding used in the 10 base-T Ethernet standard for baseband transfer*

*In Manchester coding the periods used for transferring a bit are split up. The first half contains the complementary value and the second half contains the value to be represented. This is a relatively simple way of implementing the transfer of voltage-free signals whose timing recovery from the datastream is independent of the information that is transferred.*

The transfer properties of electro-magnetic waves through the air mean that it is impossible to transfer digital signals in the baseband in the way used in many standards for high-performance wired networks. As a result, for example, the signals in an Ethernet network are converted in a coding process and sometimes even combined with each other so that several bits can be transferred in one symbol. However, the signals are never modulated, no matter what speed they are transmitted at. This step is not even used for data rates as high as 10 Gbps (also known as "10G-base Ethernet").

When electro-magnetic waves are used to send analogue signals, as is the case for radio or television transmitters, the audio and video signals are usually transferred as part of an amplitude or frequency modulation process. The technologies described below can be used to transfer digital signals. However, to do this the following requirements must be taken into consideration in the case of local networks [Burkley 1994]:

- The spectral efficiency, which is calculated in bits/MHz/area, must be as high as possible in order to achieve the greatest possible number of channels with maximum bandwidth. This is a core requirement because the available frequency spectrum is usually very limited. Please note that this aspect can be formulated as the demand for maximum spatial repetition frequency. This is required so that modulation technology can be used in cellular networks in which neighbouring cells use the same modulation technology with different parameters.

- The probability of transmission errors must be as low as possible. This requirement can also be regarded as a requirement for the minimum probability of bit errors, the *Bit Error Rate* (BER).

- The time and effort involved in implementing the modulation technology for both transmitters and receivers must also be as minimal as possible.

The requirements described here fundamentally conflict with each others' targets (see Figure 2.10). As a result the best possible implementation depends on which target function you select. The following examples illustrate this:

*Figure 2.10:*
*Target conflicts*
*in selecting the*
*type of*
*modulation*

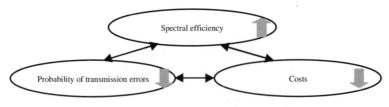

- The spectral efficiency of modulation technology can be increased by grouping several bits into one symbol which can then be transmitted via more than two permitted values. This only slightly improves the *Signal-to-Noise Ratio* (SNR). The probability of bit errors increases accordingly. The next sections describe transfers using several possible values (*MultiLevel Signalling* – MLS).

- In many cases there are ways of improving the quality of a receiver by expending more effort. In particular, adaptive reception filters that can also equalise the temporally-variable properties of the channel achieve significantly better results, measured as a reduced bit error rate. On the other hand, each additional method increases costs, usually in both the development and the production stages.

The modulation procedures shown below are based on the fact that the amplitude A(t), or the phase position *φ*(t) or the frequency *ω*(t) is used to change one of the three variable parameters in equation (2.1), depending on the information to be transmitted.

$$s(t) = A(t) * \cos(\omega(t) * t + \varphi(t)) \tag{2.1}$$

Here the spread spectrum techniques are the first real expansion of this method. However technical implementation can still be carried out in a similar way.

Because amplitude modulation is relatively prone to interference it has lost much of its popularity and nowadays is hardly used at all in digital transmission systems. For this reason it is not discussed in any more detail.

## 2.5.2    Phase shift keying

*Phase Shift Keying* (PSK) involves changing the phase of a signal that is being transmitted without changing its frequency or amplitude. If only two different phase positions are available for the transmission of a binary signal, this **B***inary* **PSK** *modulation* (BPSK) can be used to represent the signal in the following way:

$$s(t) = b(t) * A * \cos(\omega t) \tag{2.2}$$

where b(t) = +1, when a logical one is transferred, and

b(t) = –1, when a logical null is transferred.

On the receiver side demodulation can be carried out using a similar set-up to the one on the transmitter side. If we include the noise signals that are added to the useable signal in the channel, a more time-consuming procedure such as coherent demodulation, as shown in Figure 2.11, must be used instead.

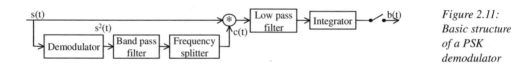

*Figure 2.11: Basic structure of a PSK demodulator*

Starting with the received signal

$$s(t) = \pm A * \cos(\omega t + \vartheta) \tag{2.3}$$

with the phase delay $\vartheta$ between the transmitter and receiver signal, the signal is squared. As a result, regardless of the prefix, and therefore regardless of the status of the transferred bit, you get

$$s^2(t) = A^2 * \cos^2(\omega t + \vartheta) = \frac{A^2}{2}(1 + \cos(2(\omega t + \vartheta))) \tag{2.4}$$

If this signal is now passed to a frequency and amplitude divider, you obtain the carrier $c(t)$. If you then multiply this by the received signal $s(t)$ you get:

$$s(t) * c(t) = \pm A * \cos(\omega t + \vartheta) * \cos(\omega t + \vartheta) = \pm \frac{A}{2}(1 + \cos(2(\omega t + \vartheta))) \tag{2.5}$$

This signal is then passed to an integrator which evaluates the prefix for its contents, at the end of each bit interval.

However, in the case of a multi-level transfer (*MultiLevel Signalling* – MLS), more than two possible states can be accepted. If there are four possible states, this is called QPSK (*Quaternary PSK*). If even more values are available the modulation is called *M-value PSK* (M-ary PSK).The phase positions can then be illustrated in diagrams, as shown in Figure 2.12 for transfers in IEEE802.11-compliant systems.

PSK is also quite often used with amplitude modulation as part of a hybrid modulation system. This is then known as *Quadrature Amplitude Modulation* (QAM). In it, the normal and quadrature components of carrier variation are modulated by independent data sequences.

*Figure 2.12: Phase diagrams for PSK and QAM modulation. IEEE802.11-compliant systems used as an example*

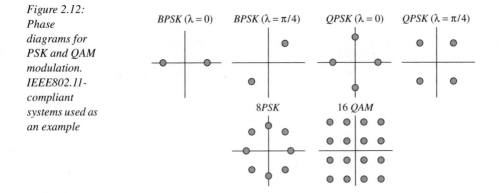

### 2.5.3 Frequency or phase shift keying

In phase shift keying, also known as *Frequency Shift Keying* (FSK), the carrier frequency is modified according to the binary signals. This could be written in the following way, in terms of equation (2.1)

$$\omega(t) = \begin{cases} \omega_0 & \text{"0"} \\ \omega_1 & \text{"1"} \end{cases} for$$

(2.6)

When selecting the frequencies it is important to ensure that the bandwidth is not significantly increased. However, there must be enough difference between the frequencies that demodulation can be carried out without involving an unreasonable amount of effort. This is especially easy to see when you look at the structure of a non-coherent receiver as shown in Figure 2.13: the demands put on the band pass filter increase as the interval between the two frequencies gets smaller.

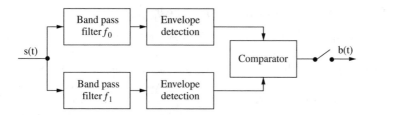

*Figure 2.13: Basic structure of an FSK demodulator*

In addition there should be as few phase shifts at the signal junctions as possible so that the simplest possible output amplifier can be used. *Minimum Shift Keying* (MSK) is a very good way of meeting these requirements. In this process a phase difference of 90 is achieved during the transfer of each bit, resulting in a signal with no shifts.

The spectral efficiency of an MSK system can be improved by filtering the digital bit stream. Often this kind of Gaussian filter is used:

$$H(\omega) = e^{-0,45\omega/\omega_c}$$

If this shows a bandwidth time product (*Bandwidth Time– BT*) of greater than one, an MSK modulated signal will still be obtained. However if the BT product is less than one (BT < 1), the symbols overlap (*Intersymbol Interference –* ISI). This permits a more robust attitude to noise but requires a more complex demodulator.

# 2.6 Channel access

## 2.6.1 Overview

Before several channels can operate independently of each other, in particular several wireless systems, the shared channel needs to be split up. Four basic processes are used to do this:

- *Time Division Multiple Access* (TDMA)
- *Frequency Division Multiple Access* (FDMA)
- *Space Division Multiple Access* (SDMA)
- *Code Division Multiple Access* (CDMA)

These are all covered in the description of *Spread Spectrum Techniques* (SST) in section 2.7.

The *Orthogonal Frequency Division Multiplex* procedure (OFDM) which is based on FDMN (see section 2.8) makes it possible for several sub-frequencies to use one bundled channel.

## 2.6.2 Time division multiple access

*Time Division Multiple Access* (TDMA) is based on the fact that the communications elements belonging to different communications participants can be transferred one after the other over the same channel. If this time sequence involves the output and feedback channel between two communications partners, it is also known as *Time Division Duplex* (TDD). In this case the time division multiple access procedure uses a high-bit-rate transfer channel, which is distributed among several channels with a lower data rate during the time sequence. The central feature of time division multiple access is that it is relatively easy to implement a central station. It is especially easy to implement a frequency mixer. Supporting handover by changing the base station also appears to be equally uncomplicated because it does not interrupt the transfer of the data stream. The data rates of the individual communications participants can also be increased by simply extending the duration of the time slots. In real life this type of time slot bundling is carried out in packet-oriented transfers as part of the *General Packet Radio Services* technique (GPRS) in *Global System for Mobile Communication* systems (GSM) and for data transfer in the DPRS technology used in DECT systems, as described in section 6.5.2. However, the timeslots for the various participants are usually managed centrally. One disadvantage is that a TDMA channel needs relatively high bit rates for time division to be practical and effective. Synchronisation between the transmitter and receiver is also somewhat more time-consuming because the transmission between the two stations is constantly interrupted.

## 2.6.3 Frequency multiplex procedure

*Frequency Division Multiple Access* (FDMA) supplies a frequency range for a communications stream that is independent of the frequency ranges used by the other communications participants. It is also comparatively easy to allow any station that is ready for communications to select its own frequency alongside a centralised frequency management procedure. To do this a station can, for example, check the field strength of all possible transfer frequencies and assume that the transfer frequency with the lowest field strength is free.

### 2.6.4    Combined procedure

Combinations of both procedures can also be used, increasing the number of possible users. For example, DECT systems use a combination of TDD, TDMA and FDMA procedures with twenty time slots in which every ten time slots are assigned a direction of transfer, and twelve frequencies (see Figure 6.1). However, the QAM modulation described in section 2.5.2 illustrates such a combination of amplitude and phase modulation.

### 2.6.5    Space division multiple access

In *Space Division Multiple Access* (SDMA) the channels used for transmission are used time and time again at specific geometric intervals as part of what is known as "cluster formation". In this way an unlimited amount of traffic can be transferred in an unlimited amount of space, despite a limited number of channels. This procedure is based on the fact that the field strength of a radio signal becomes lower the further it gets from the transmitter because of the attenuation effects described in section 2.4.3. If two transmitters are far enough apart their signals are so weak that they cause no noticeable interference to each other.

## 2.7    Spread spectrum techniques

### 2.7.1    Basics

The procedures described in section 2.5 have also been widely used in real life systems, due partly to what was technologically achievable, historically. Despite this they do have several disadvantages that are particularly noticeable in the operation of mobile radio networks. However, the spread spectrum techniques described below can be used to compensate for these disadvantages [Hatzold 2000].

The standard procedures for transferring messages with digital modulated HF carriers use a comparatively narrow bandwidth. In ideal band limitation the bandwidth equals the symbol rate but, in real life, because of the edge steepness (*roll-off* characteristic) of the filter, it is usually necessary to select a bandwidth that is greater than the symbol rate. *Spread Spectrum Techniques* (SST) transfer signals with a higher bandwidth than would be needed for the symbol rate alone.

As a result this procedure is very sensitive to variations in field strength caused by changes in reflection conditions along the diffusion path (*fading*). The location of the receiving antenna may cause a great variation in the service that can be received. As the spatial characteristic of the vertical waves that are created in this way depends on the actual wave lengths, the greater bandwidth may reduce the local variation in field strength.

According to Shannon the transfer capacity $C$ in bits/s of a message channel can be calculated by

$$C = B * ld(1 + S/N)$$

(2.7)

Here, $B$ is the channel's bandwidth and $S/N$ is the signal-to-noise ratio (SNR). Therefore, if you increase the bandwidth, the channel capacity also increases linearly. However, if you use this method on its own it will conflict with the requirement for maximum spectral efficiency as described in section 2.5.2.

Channel capacity will only increase logarithmically with the power of the transmitter.

An important prerequisite for efficient usage (i.e. for maximum possible transfer capacity within a limited frequency range) is the selection of a procedure in which error-free overlaying of the frequency bands is possible on different transmission routes.

Spread spectrum techniques also help make transmission much less sensitive to frequency-selective interference. The historic origins of this procedure can also be seen here. The original impetus for developing spread spectrum techniques dates back to the development of military radio systems, for which stability when faced with interference from enemy transmitters was a crucial factor. These usually had a high signal strength in a limited frequency bandwidth. With spread spectrum techniques it is also possible to simultaneously operate many autonomous stations in a particularly noisy environment effectively because two effects are achieved at the same time:

• a high degree of insensitivity to interferences can be achieved on the transfer channel.

• at the same time a station operating with spread spectrum techniques also has a comparatively low impact on its environment.

This is why spread spectrum techniques are so well-suited for use in environments with high levels of interference, or for a large number of autonomous stations. For example, the use of spread spectrum techniques in the 2.4 GHz ISM band is linked with a range of specific framework conditions.

There are two basic procedures: the *Frequency Hopping Spread Spectrum* (FHSS) procedure and the *Direct Sequence Spread Spectrum* (DSSS) procedure, which are described in sections 2.7.2 and 2.7.3. Section 10.3 lists additional criteria that restrict both these technologies.

## 2.7.2    Frequency hopping spread spectrum procedure

The principle behind the frequency hopping spread spectrum procedure is that both the transmitter and the receiver change the carrier frequency in accordance with a *Pseudo Random Sequence* (PRS) of numbers. This change of carrier frequency is called a *hop*. Figure 2.14 shows the basic procedure. For example, transmitter 0 has agreed frequency sequence $f_1, f_{76}, f_2, f_{78}, f_3, f_{75}, f_{76}, f_{77}, f_4,$ $f_{79}, f_{78}, f_{77}$ with its receiver. This means that any interference in the transfer from Station 1 by Station 3 transmitting continuously on one frequency, which may be intended as radio jamming, or by another FHSS Station 2, which has defined a different frequency sequence with its communications partner, are limited to short periods of time.

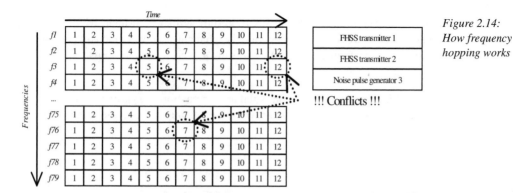

*Figure 2.14: How frequency hopping works*

There are two types of FHSS transfer:

- fast FHSS transfer in which only one symbol is transferred between two frequency hops, and
- slow FHSS transfer, in which several symbols are transferred on each frequency hop.

With regard to system architecture, there are

- asynchronous or uncoordinated FHSS systems. In these systems the frequency hops are not co-ordinated and therefore occur on different transmission routes and even in overlapping radio ranges.
- synchronous or co-ordinated FHSS systems in which the frequency hops from all stations occur simultaneously within one radio cell.

The most important advantage of the frequency hopping spread spectrum procedure is that it is relatively easy and therefore inexpensive to implement the appropriate systems. Most of them also manage to be energy-efficient.

## 2.7.3    Direct sequence spread spectrum process

In the DSSS procedure the frequency spread is achieved by a logical exclusive/or

(XOR) operation on the data with a *Pseudo-Random Numerical Sequence* (a PRS or PN sequence), which results in a higher bit rate. The individual signals within this PN sequence are called "chips". Figure 2.15 shows the structure of this operation at the transmitter end. This is inverted at the receiver transmitter end. You will find more details on this process in Appendix A.2.

*Figure 2.15:*
*Basic structure*
*of a DSSS*
*modulator*
*([Zyren 1])*

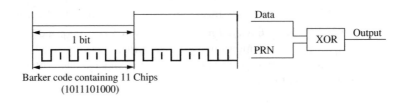

Barker code containing 11 Chips
(1011101000)

The fundamental advantage of this procedure is that high-intensity narrowband interference $I$ are spread throughout the channel as lower intensity broadband noise $I_{PN}$ (see Figure 2.16). As a result, a high level of insensitivity can be achieved.

The DSSS procedure can be compared to the situation at a party, in which you can always hear key words, such as your own name, against the background of general noise.

*Figure 2.16:*
*Changes in the*
*frequency*
*spectrum*
*during the*
*DSSS*
*procedure*

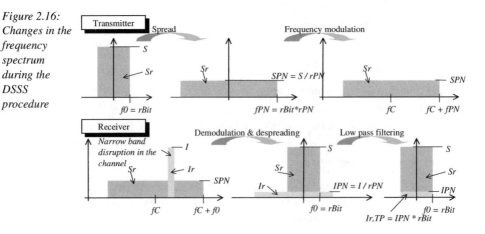

The DSSS procedure involves the following features:

- The DSSS procedure has a high degree of spectral efficiency and can therefore achieve high user density in a limited bandwidth and with high data rates.

- The transfer rate increases if the useable data is linked with the higher bit rate PN code. This not only increases the speed requirements placed on the transmitter module, but also the amount of power used.

- At the receiver end several receivers are then required to evaluate the signal. This in turn results in a more complex receiver module and an increase in energy consumption.

## 2.8    Orthogonal frequency division multiplex procedure

*Orthogonal Frequency Division Multiplex* (OFDM) procedures are based on parallel data transfer using frequency multiplexing. They are also known as *multi-carrier* or *discrete multitone* procedures. They permit high data transfer rates with comparatively good resistance to interference, particularly against narrow band interference. The data is split among a number of transfer channels which ensures that only a very small number of bits will be transferred incorrectly. These can then be corrected by using the appropriate encoding. An added benefit is that each sub-frequency can be efficiently modified to suit the properties of each transfer channel. This is a good way to counter the difficulties that are particularly associated with variations in signal runtime (*delay spread*) caused by using different spread paths (*multi-path*). Wireless transfer in frequency ranges of 5 GHz are especially prone to these problems. OFDM is also used alongside wireless systems for digital audio signals (*Digital Audio Broadcasting* – DAB) . OFDM is even used with wired transmission protocols with high spectral efficiency levels such as *Asymmetric Digital Subscriber Line* (ADSL).

Unlike the usual frequency division multiplex procedures, which have relative large distances between the various channels and therefore only use a small part of the frequency spectrum, the different channels used in OFDM procedures are allowed to overlap: a fundamental difference. Nevertheless OFDM uses *subcarrier frequencies* to avoid interference on the channels. This means that the signal rate $T$ of each subchannel is specifically selected to match a frequency interval between two adjacent subchannels.

Figure 2.17 shows the Fourier transforms of pulse-shaped signals. These pulses, whose width is $T = 1/f_0$, are multiplied by a carrier frequency $f_n$, producing the transformed signals $si(f\text{-}f_n)$. This has null spaces for all frequencies to which the following applies:

$$f_{Null} = f_n \pm N * f_0 \text{ with } N = -\infty,...,-2,-1,0,1,2,...,\infty \qquad (2.8)$$

If we now use carrier frequencies that are shifted by $f_0$, then subchannels can be received independently of each other. In this context "independently" means that none of the distortions occur that are usually caused by the overlapping of carrier frequencies due to *Intersymbol Interference* (ISI). In this case the technical prerequisite is that the carrier frequencies are extremely precise.

*Figure 2.17:*
*Illustration of*
*orthogonal*
*subcarrier*
*frequencies in*
*an OFDM*
*transfer in the*
*frequency*
*range. The*
*other carriers*
*have a null at*
*the maximum*
*point of a sub-*
*frequency*

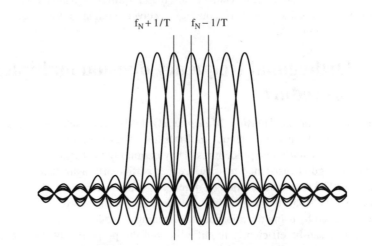

*Figure 2.17:*
*Illustration of*
*orthogonal*
*subcarrier*
*frequencies in*
*an OFDM*
*transfer in the*
*frequency*
*range. The*
*other carriers*
*have a null at*
*the maximum*
*point of a sub-*
*frequency*

Figure 2.18 shows examples of three carrier frequencies that can be demodulated independently of each other.

*Figure 2.18:*
*Three example*
*orthogonal*
*subcarrier*
*frequencies in*
*an OFDM*
*transfer in a*
*time range.*
*To make things*
*clearer, the*
*three carrier*
*signals have*
*the same*
*amplitude and*
*same phase*
*position*

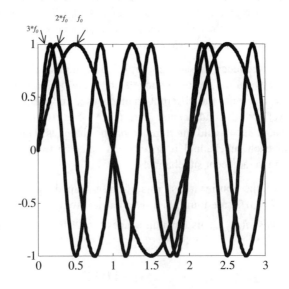

Nonetheless, OFDM receivers can be implemented very cost-effectively because they do not require a steep edge filter or an *equaliser*. You simply need to carry out a *Fast Fourier Transform* (FFT) or its inverse function, an *Inverse FFT* (IFFT) [Walke 2000].

# 2.9 Antennae

## 2.9.1 Isotropic and anisotropic antennae

To transfer data using electro-magnetic waves you will need transmitter and receiver antennae. You must also take into consideration a multitude of influencing factors which can significantly affect your system's performance. However none of the standards described in this book set out the requirements for antenna properties.

As section 2.4.2 described, the fundamental parameter that defines the range of electro-magnetic waves is their power. In free space the field strength of an electro-magnetic wave reduces proportionally to the distance from the transmitter. Accordingly, the input strength at the receiver reduces by the square of the distance.

If we now start from the ideal point-shaped emission source shown in Figure 2.19, whose transmission $P_S$ is equally distributed in all directions, i.e. omnidirectional, this transmission spreads out in a spherical formation in the air.

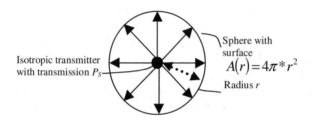

*Figure 2.19: Emission characteristics of an isotropic antenna*

The entire beam is therefore distributed at a distance $r$ from the transmitter on a spherical surface with $A(r)$. The beam density is received by a pointed receiver antenna located on the sphere's surface.

$$p = \frac{P_S}{4\pi * r^2} \quad \left[\frac{W}{m^2}\right]$$

(2.9)

In real life antennae are used that tailor their performance to suit specific needs. These antennae are described in more detail in Appendix A.3.

## 2.9.2 Diversity

You can compensate for the extreme sensitivity of the spatial positioning of a receiving antenna by using several antennae that are physically separate. To make this easier to set up, these antennae are supplied in one case and typically cover gaps of less than half a wave length. In this way you can use *antenna diversity*, i.e. multiplicity or variety, to select the specific antenna that lies in the range of greatest field strength. You can also bundle the directions by using the interference effects of the different antenna sectors. Depending on the electronic switching used, this can also allow you to set a particular receiver antenna to the direction of the strongest signal.

### 2.9.3   Polarisation

The polarisation direction of the electro-magnetic wave also plays an important role. In rod-shaped antennae, in particular, the signal is emitted with a preferred polarisation direction which a rod-shaped receiver antenna also expects to receive in order to achieve the best possible performance of the charge carrier in the antenna. In many cases the selection of polarisation direction is therefore defined by the site and the type of antenna construction. The implementation of printed board antennae in PC adapter cards, as used in laptop computers, requires a horizontal polarisation. In the best case this polarisation is also used by all the communications participants involved [Andren 2001].

### 2.9.4   Location

In contrast, the directional characteristics of many mobile applications can cause interference. In this case you should try and achieve equal i.e. isotropic emission in all directions. Although it is at least generally possible to achieve a high level of directional independence in antennae used over great distances, in real life you must also take into account shadowing caused by obstacles. This is why the location of your antenna is of crucial importance. However, antennae are so small that they can be installed in a number of ways. An antenna that is as long as a wave length in the 2.4 GHz range measures only about 12.5 cm. Correspondingly, a $\lambda/4$ antenna can be reduced to little more than 3 cm. There are two basic installation types:

- Integrated into a device. A multitude of different construction types are used for this purpose:

  - In the PC adapter card of a laptop computer with no external adapter. This limits the direct direction of emission to a sphere from the PC adapter card slot.

  - In the PC adapter card of a laptop computer with an external adapter. This increases the direct sphere of emission. However, there still remains a strong radio shadow on the side that the signal must first cross before it reaches the computer.

  - In the laptop lid. This technology provides the best results and is the most user-friendly, as long as the lid does not contain any metal parts. However, in this case the antenna must usually be installed by the laptop manufacturer. Therefore antennae of this kind will only be available in the more expensive devices for the foreseeable future.

  - In a desktop PC. This construction is not often used because PCs are often placed in unsuitable locations under the desk, which can cause a shadow, and also because some PCs have a metal frame.

- Using an external antenna. Although this connection introduces another peripheral device to the tangle of cables on the desk and is not especially handy for highly-mobile devices, it does give the best results for devices that are usually stationary because radio results get significantly worse if a laptop is moved even slightly. In particular the uncomplicated USB connection appears to be growing in popularity.

### 2.9.5    Selection

Antennae come in an almost unlimited number of shapes and sizes. As they can be used in active devices no matter what the protocol or manufacturer, Figure 2.20 provides an overview of the basic parameters, with a few examples for each one that you should take into account when you select an antenna. The thing to remember here is to arrange the attenuation of the antenna connection when you install it because transmission takes place directly in the RF range. The attenuation must be at least 0.2dB/m.

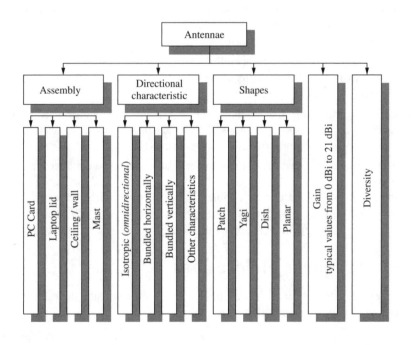

*Figure 2.20: Overview of the most important technical parameters for selecting an antenna*

## 2.10   Special features of wireless networks

### 2.10.1   Wireless networks use a common medium

The wireless transfer of information has a number of special features which do not affect wired systems (or only to a very limited extent). The crucial point is that the transfer takes place through the air, which is therefore a common medium. This is true in two respects:

- Different stations access the same medium, and can therefore mutually influence each other. If they use the same transfer channel it is vital that they comply with specific regulations for accessing the medium.
- The regulations governing the use of "air interfaces" are issued by a range of authorities to ensure that general commercial interests are represented.

### 2.10.2  Spatial characteristics

The installation of transmitters and receivers in a particular space leads to various effects that play an important role in the design of wireless networks. As the signals are distributed spatially it is not possible to monitor the channels from one location. In particular it is not possible for the transmitter to use collision identification, as supported in a classic Ethernet architecture. This problem is known as the *Near-Far* or *Hidden Station Problem*. Also, the signal's spread is unrestricted. However, a signal's strength reduces in relationship to the effects of attenuation, as described in section 2.4.2. This has four particular effects:

- The relationship between the strength of the signal and the interference (*Signal-to-Noise-Ratio* – SNR) is only large enough for satisfactory reception quality within a specific distance.
- If several radio cells are in operation it may happen that one device receives signals from several radio cells at the same time. The device must then decide which radio cell is active at that time.
- It is not impossible that signals sent within the radio range may also be received by unauthorised stations. To prevent unauthorised "eavesdropping" you will need the appropriate security measures.
- The spatial characteristics of the transfer channel may change over time. This is why you cannot modify the channel as you do for wired media to suppress the reflections at the end of a cable. In the same way, transferred messages may not only be destroyed by the overlapping of two different message packets but also by the overlapping of the original and the reflected signals.

## 2.11  Frequency allocations

The technologies described in this book can be divided into three categories according to their frequency allocation:

- A frequency range that is reserved for use by a specific radio technology. For example, the 1880 MHz to 1900 MHz range which is reserved in the area governed by ETSI for DECT technology (see Chapter 6).
- A frequency range that is reserved for use by specific radio technologies that comply with a particular set of regulations. The *Unlicensed National Information Infrastructure* (UNII) band (used in the USA) or the European *license-exempt frequency band* in the 5 GHz range are examples of this.
- A frequency range that is reserved for use by a wide variety of applications that emit electro-magnetic waves. The license-free ISM (*Industrial, Scientific,*

*Medical*) bands are particularly important in this context because they are used throughout the world for industrial, scientific and medical purposes. The frequencies in the 2.4 GHz range in particular are used by numerous communications protocols such as IEEE802.11, Bluetooth and HomeRF, and may later be used by Upbanded DECT. However, other transmitters (such as microwave ovens!) also use this range. They are the main reason that this frequency range has been released because it would otherwise cause too much interference for other transfer procedures that are licensed, and for which users pay a fee.

However, in this context it must be mentioned that this 2.4 GHz band has not been completely released in every country. This results in a lower number of channels. Japan, France and Spain are particularly affected by these restrictions. However, since the beginning of the year 2000 the entire spectrum from 2400 to 2483 MHz has been made available in France and Spain. There are also clear differences between the other specifications, concerning issues such as permitted emission levels and modulation procedures.

In North America and Europe you must implement a pre-defined frequency spread procedure before you can use the 2.4 GHz ISM frequency range.

If you use an FHSS procedure you can only use a channel whose frequency is less than 400 ms. Two frequencies that follow one another must have an interval of 6 MHz.

If you use a DHSS procedure the regulations for its use originally specified a PN code with a length of at least 11 bits.

Figure 2.21 shows an overview of the frequency allocations in Europe. Regional variations are discussed in the relevant sections.

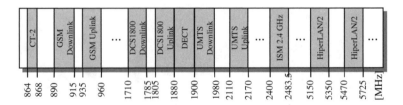

*Figure 2.21: Overview of frequency allocations in Europe*

# 3

# Applications, devices and standards

## 3.1 Application scenarios

Initially wireless information transfer only replaced cable in the Physical layer in the ISO-OSI reference model. However, this basic statement needs to be expanded:

- The wireless networks adopted the main network architectures, which had established themselves over the decades of using wired networks, for use. As a result the structure of popularly-used network applications and the application scenarios described in the sections below have a great deal in common.

- The various different mobility features offered by the use of wireless networks can make the integration of intelligent value-added services look very attractive. However, if these services in the higher layers are connected with a physical transmission protocol in the lowest layer, as is the case with Bluetooth, the actual wireless network is still more than a mere replacement for cable.

### 3.1.1 Personal Area Networks

The relatively new term *Personal Area Network* (PAN) is used to describe communications between the devices of one or a few more users within a range of around 10 m. There are three different scenarios to mention here:

- The connection of peripheral devices such as printers, mobile telephones, *Personal Digital Assistants* (PDAs), or digital cameras to a PC to transfer or synchronise data.

- The connection of external user devices with the service platform. A typical example was the early innovation of the Ericsson headset for mobile telephones based on Bluetooth.

- The connection of several PCs for data transfer. This architecture directly borders on the functionality of classic LANs.

Taking the applications as a starting point we can assume that the communicating devices are usually in the same space or at least quite close to each other. Most applications will therefore function quite happily with moderate bandwidths.

However, for these PAN applications it is essential that the radio modules can be produced as cheaply as possible because the functionality must also be implemented in relatively simple and inexpensive devices to achieve the required levels of acceptance.

In this case it is vital that more than one point-to-point connection can be supported and that data rates that allow an effective connection to a range of more powerful devices can be achieved. This is why radio mice and radio keyboards, as supplied by various manufacturers for use in the 27 MHz band [Gieselmann 2001], are not included among PANs. Moreover, it is very probable that PAN technologies will replace these proprietary systems.

### 3.1.2    Local Area Networks

A *Local Area Network* (LAN) involves the connection of several devices in the same area of a building. A typical scenario is the linking together of client PCs, servers and peripheral devices, such as printers, in an office building. The typical distances between these devices (stations) may be a few tens or hundreds of metres. The bandwidth requirements are much higher than for a PAN, as it must also be possible to exchange larger files and even operate applications over the network. In real life these networks use three typical application classes.

### 3.1.3    SoHo LANs

Many smaller offices and households (*Small Offices and Home Offices – SoHos*) have a small number of computers that are networked together and can access common resources such as a printer or a modem. For wireless systems in particular this scenario means:

*   Modest system performance requirements as they usually only involve small quantities of data and the number of stations in the network is very limited. One radio cell is usually sufficient to supply the various computers.

*   Costs are of crucial importance, as investments are in most cases financed from the users' own pocket. On the other hand the downward pressure on price is not as high as for PANs, as the number of required modules is usually very low and the complexity of the equipped systems is comparatively high.

*   The number of users and stations is limited. As a result, only very basic tools are needed for network administration or sometimes even none at all. As far as possible, these tools must be intuitive to use so that even users who are not IT professionals can set up and manage these systems.

The SoHo market represents a vital aspect of market penetration for WLAN system manufacturers for two reasons: potentially, very high unit sales can be achieved in this market segment, so the manufacturers can benefit from effects of scale, and, on the other hand, the technical knowledge needed to operate these units is usually not as high as in larger office environments so the users' learning curve is not so steep [Frost & Sullivan 1999].

### 3.1.4    Office LANs

In contrast, the connection of computers in larger offices is placing ever-increasing demands on networks. The connection of hundreds, or even several thousands, of computers at one site is no longer a rare occurrence.

However the following points must be taken into consideration when wireless systems are used:

- Demands on system performance are considerably higher in these environments than in SoHo environments. In corporate office environments, many times more data is exchanged because the number of stations in the network can be extremely high. In many cases one radio cell is not enough to achieve the physical coverage or the required number of channels.

- Costs are of crucial importance as the number of modules required is often comparatively high.

- There may be a very large number of users and stations. As a result, powerful tools, usually operated by experts, are needed for network administration in these environments.

### 3.1.5   Connecting networks

It is also necessary to include the connection of networks because wireless networks are often used for this purpose:

- *LAN-LAN connection:* The definition of a local network reaches its limits when the devices are no longer located in one building but are distributed throughout several locations. In many cases the aim is to link the networks directly to each other without taking the step of transferring data via a long-range communications network. However, for the user the transparent interconnection of LANs can be achieved by using a virtual network, even over a long-range communications network.

- *LAN-WAN connection:* Another area of overlap is the connection of local networks to long-range communications networks in the sense that LAN technology is used to create physical connections locally. In this case, for wireless transfer systems, we refer to the *Wireless Local Loop* (WLL).

## 3.2   Device types

A variety of devices must be provided to fulfil these tasks, in particular:

- network adapters that link end devices. These can be divided into five categories:
  - PC cards for use in laptops.
  - PC cards for use in desktop computers. As the connection of stationary desktop PCs is less important for the market than the wireless networking of portable laptops, PCI solutions are often based on PC adapters. Compatibility is achieved with the help of additional circuits.
  - USB stations for flexible use in desktop and laptop computers. However, both the PC hardware and the operating system must support this. This is not the case either for older PCs or for more recent computers that use Microsoft Windows NT 4.0.

- integrated WLAN adapters installed by the manufacturer in a laptop.
- network adapters used to link embedded systems. Until now these have often still been application- or customer-specific solutions.
- access points that create the connection between wireless segments and wired segments in a network.
- bridges to create a LAN-LAN or LAN-WAN connection.

## 3.3   Standards

At the time of publication, the following candidates appear to be the most significant, of all the products available on the market:

- IEEE802.11
- Bluetooth
- DECT
- HomeRF and
- HiperLAN/2

They are described in more detail in Chapters 4 to 8. Figure 3.1 contains an overview of the two most important parameters:

*Figure 3.1: Overview of WLAN technologies. The IrDA infrared technologies, and access technologies using mobile telephony networks, are also listed for comparison*

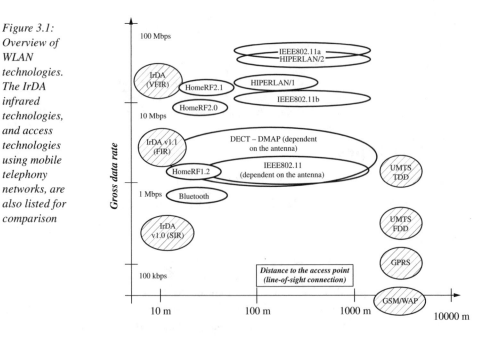

This figure requires a few extra comments for two reasons, already mentioned:

- The bandwidth shown is the gross bandwidth. The actually achievable net data rate is often much lower. The difference is due to unused time slots, collisions and even the protocol itself in the sense that a header for addressing and qualification must be added to the packets and also that the stations still exchange information for traffic control between themselves. In some situations, these rates can differ widely from one system to another. On the other hand, the values also depend greatly on the associated traffic levels.

- The range refers to the unrestricted propagation when a direct *Line Of Sight* (LOS) is present. The large ranges shown in the figure result from the use of directional antennae which, without any change in transmitter power, can increase significantly in comparison with an isotropic antenna. However, this aspect is pretty much independent of the protocol.

In real life, however, a line of sight is rarely available. Additional obstacles increase the damping (attenuation) effect. This either reduces the range or the achievable bandwidth.

# 4     IEEE802.11

## 4.1     The standard

### 4.1.1     Positioning

For *Local Area Networks* (LANs), the IEEE's 802.x family of standards (see also section 1.5) forms an all-embracing platform. Especially important members of this family include the Ethernet standards, based on 802.3, in the different speed classes from 10 Mbps to the current top speed in the specification, 10 Gbps. Figure 4.1 shows an overview of the main standards and how they interrelate.

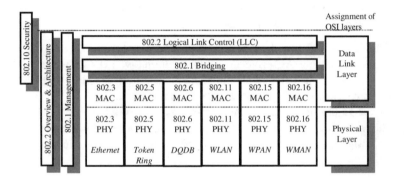

*Figure 4.1: The most important standards in the IEEE 802 family and their relationships*

In 1997, after a seven year development and approval process, the IEEE's *Standards Association* (SA) agreed on the first non-manufacturer-specific standard for transfer protocols in wireless local area networks. This is why the original standard is sometimes referred to as IEEE802.11-1997. Since April 2001 the 1999 edition of this standard can be downloaded from the Internet (*http://standards.IEEE.org/getieee802/*). You will find a general overview of the IEEE802.11 technologies in [Sikora (1)]

Other committees have been set up within the framework of the IEEE802.11 standard. Their aim is to ensure the market success and interoperability of different devices:

- The *Wireless LAN Association* (WLANA) (*http://www.wlana.com*) has the task of supporting the acceptance of the standard through marketing and information activities.

- The *Wireless Ethernet Compatibility Alliance* (WECA), *http://www.wi-fi.com*, founded in August 1999, certifies the interoperability of 802.11-compatible devices. Incidentally, devices that comply with IEEE802.11 are also marketed under the brand name Wi-Fi™ (*Wireless Fidelity*). Prior to this the *Wireless LAN Interoperability Forum* (WLIF), founded in March 1996, followed a similar aim. This committee also produced a specification concerning interoperability checks which was published in April 1997.

## 4.1.2    Structure

The 802 standard family uses the terms *Physical Layer* and *Data Link Layer* to name the lowest two layers of the ISO/OSI reference model (see also section 2.2.1). In this model the connection layer is further sub-divided: access control is provided by *Medium Access Control* (MAC) and *Logical Link Control* (LLC), which is identical in all standards that comply with IEEE802, and provides logical control of connections. This allows protocols from the upper layers to access communications services regardless of the access mechanism used, and their physical implementation. In particular, this means that wireless protocols that comply with 802.11 in protocols such as TCP/IP (layers 4 and 3) can, in principle, be used in exactly the same way as the more usual wired 802.3 protocols. However, the multiple implementation of algorithms to correct errors or control traffic may still lead to bandwidth restrictions and therefore a reduced data rate. This is explained in greater detail in section 10.5.

The IEEE802.11-1997 standard specifies data rates of 1 and 2 Mbps. The standard describes one MAC protocol and three alternative PHY (physical-layer) protocols. The terms "frequency hopping spread spectrum" (FHSS) and "direct sequence spread spectrum" procedure (DSSS) are used to describe two spread spectrum techniques on the basis of electro-magnetic waves in the 2.4 GHz ISM band. This standard also proposed an infrared technology. However, this has never been used in a real life scenario and is therefore not described in this book.

*Figure 4.2:*
*Assignment of*
*layers to the*
*IEEE802.11*
*standards*

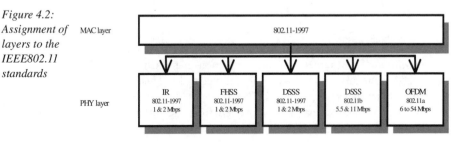

## 4.1.3    Extensions

In 1999 two more elements, IEEE802.11a and IEEE802.11b, were added with the specific aim of achieving higher bandwidths. These two elements use very different methods to do this. 802.11a uses an additional frequency range in the 5 GHz range to provide higher bandwidths. In contrast 802.11b continues to use the 2.4 GHz ISM band. The current version of 802.11b can achieve a gross data rate of 11 Mbps using the DSSS technique. This is based on the 802.11 DSSS PHY layer so that systems that comply with 802.11-1997 can be used with data rates of 1 or 2 Mbps. However, it is not possible to use this arrangement to connect to WLAN systems that comply with 802.11 and use the FHSS procedure. The extensions are described in section 4.7.

Both extensions are based on the same MAC protocol and can therefore be used in identical system architectures.

# 4.2    Architectures

## 4.2.1    System architectures

The 802.11 standard describes the following operating types, whose hierarchy and structure are shown in Figure 4.3 and Figure 4.4.

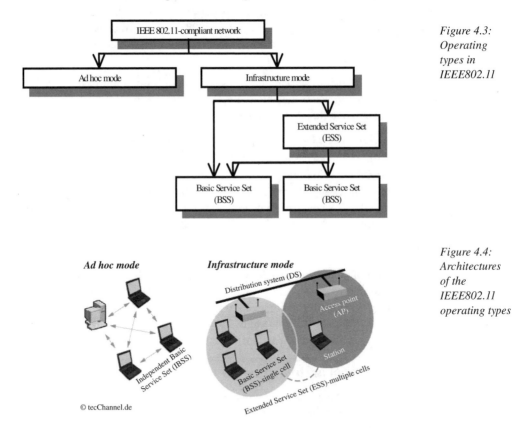

*Figure 4.3: Operating types in IEEE802.11*

*Figure 4.4: Architectures of the IEEE802.11 operating types*

In *ad hoc mode*, end devices in a *peer-to-peer* network can communicate directly with one another. In this *Independent Basic Service Set* (IBSS) you can quickly create simple and cost-effective networks, usually over short distances and with a limited number of participants, and without integration into a larger network structure.

In *infrastructure mode* communication takes place via one access point. With this operating type even stations outside the direct range can communicate with each other. A portal or a gateway can be used to provide a connection to an external network.

In its simplest version this kind of network consists of one access point and a group of wireless 802.11 stations in the access point's transmission zone. This type of network is called a ***Basic Service Set*** (BSS).

If two or more several BSSes are operated in the same network, by linking the access points via a background network, this is then called an ***Extended Service Set*** (ESS). In the standard this linking network is called a ***Distribution System*** (DS). This can be implemented either with wires, with radio links or with optical links. In ESS mode, in particular, support is required for the localisation of stations in one BSS and the changing of stations from one BSS to another. This type of change is called *roaming*. The IEEE802.11 standard does not yet include the communication between access points via the distribution network required for roaming (see also section 4.7).

In infrastructure mode, stations must log onto the access point (see also section 4.3.7). They then transfer data along the channel that is used by the access point and then assigned to them. Here, three comments are worth making:

- This assignment can be pre-defined by a system administrator to manage access rights and physically connect user groups to each other.

- However, there is another way to set up this assignment. In particular, a range of manufacturer-specific procedures are available for load balancing (see also section 9.4.3). A new assignment can also be carried out as part of *roaming*, for example, by physically changing the location of the mobile stations which results in changes to the channel characteristics.

- The description of ad hoc mode in the IEEE802.11 standard differs from the definition of ad hoc networking given in section 1.3.10. This is because, even in infrastructure mode, the access points and mobile stations must be able to exchange information with each other and execute logon independently. Therefore you should remember that IEEE802.11 networks use ad hoc networking in both ad hoc and infrastructure mode.

## 4.2.2   Protocol architecture

When we look at the protocol architecture of IEEE802.11 the feature to notice is that, in addition to the usual layer division used in other protocols (one *sublayer* and one *management layer*), there is a further division of the Physical layer (the *PHY layer*). In this division the ***Physical Medium Dependent*** (PMD) sublayer is responsible for modulation and encoding, whereas the ***Physical Layer Convergence Protocol*** (PCLP ) provides a normal PHY interface, no matter what medium is involved. In particular, PLCP also supplies the ***Clear Channel Assignment*** signal (CCA) which shows the medium's current state. In addition to monitoring signals, this CCA signal (which is transmitted by other communications participants) is also used to prevent collisions due to noise from microwave ovens. In 50 Hz power supply networks, and with an activity rate of 50%, microwave ovens transmit noise signals that last for 10 ms at 10 ms intervals. Even when the interference from a microwave oven is very high during the active phase there is still the 10 ms pause in which the WLAN can transmit [Solectek Corp. (2)].

The *Service Access Points* (SAP) for the signal flow are shown in Figure 4.5.

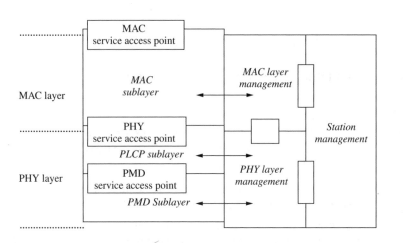

*Figure 4.5:*
*Service access*
*points for the*
*MAC and PHY*
*layers in*
*IEEE802.11*

- The MAC SAP is identical to the service access points in other standards that comply with IEEE802.x, especially Ethernet standard 802.3.
- The PHY SAP operates independently of the physical transmission itself.

Another special feature is that the data received by the two layers from higher layers is packed with an additional frame. The frame formats are described in more detail in section 4.3.6.

# 4.3    Channel access

## 4.3.1    Where it fits in

The channel access layer (MAC layer) described in the 802.11 standard is very closely related to the definition in the Ethernet standard. However, the wireless standard must also take into account the special features of its transmission routes. In particular, collision monitoring, as carried out on wired transmission media, is impossible for wireless transmissions (due to the *near/far problem*: see section 2.10.2).

Two access mechanisms, **Distributed Coordination Function** (DCF) and **Point Coordination Function** (PCF), are used for access. Here the PCF accesses DCF as shown in Figure 4.6. Below you will find descriptions of the two algorithms involved, and an explanation of the relationship between these two functions.

*Figure 4.6:*
*PCF and DCF*
*accesses and*
*classification*

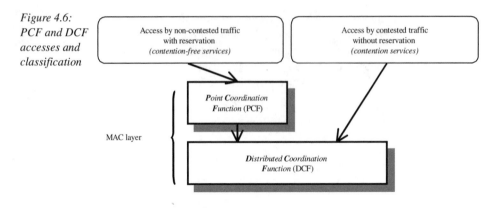

## 4.3.2    Distributed Coordination Function

### CSMA/CA

The *Distributed Coordination Function* (DCF) describes the principal method used by the IEEE802.11 algorithm standard to access the transmission channel. It is based on the CSMA/CA algorithm:

- *Multiple Access* (MA) means that several communications participants can use the same transmission channel (*shared medium*).

- *Carrier Sense* (CS) means that each communications participant can monitor the same channel and adjust their own activity to match the channel's state. In particular, no station can start transmitting if it detects that the channel is busy. Here the IEEE802.11 standard differentiates between physical listening (*physical sensing*), which uses the measured field strength to evaluate a channel's activity, and virtual listening (*virtual sensing*). A station can use a special protocol to reserve a specific channel for a particular time interval, for this purpose, as described in sections 4.3.3 and 4.3.5.

- *Collision Avoidance* (CA) describes the algorithms used to prevent collisions on the channel.

However, in this context, it must be emphasised that avoidance does not guarantee that collisions will never happen again. Even in 802.11 networks collisions can still occur that lead to data loss on the transmission channel. In the same way as for wired transfer mechanisms, the sending station is responsible for storing the data until a transfer can be carried out successfully.

So, the best way to think of the algorithms is that they are intended to reduce the probability of collisions as far as possible. Nevertheless, users should be aware that in normal operations, just like in wired Ethernet networks, collisions will still occur quite frequently and are not an exceptional event.

A communications participant must therefore first "listen" to the medium before they start their transfer. This is called *Listen Before Talk* (LBT). Two main methods are used:

- The transfer can start when the medium has been identified as not busy for a pre-defined time interval. The time interval during which the medium is physically "listened to" is called the DIFS time. The significance of this time is shown below.

- However, if the medium is identified as busy, because another station is transmitting, the transfer is put back and delayed for a particular waiting time period as part of a backoff process. No new transfer attempt can be made during this waiting time period.

This procedure is unlike the technique used in wired Ethernet networks in which the backoff algorithm is only started if a collision is detected.

**Backoff process**

The waiting time is defined as part of what is known as a backoff process. In this process a pseudo random number sequence is first defined. This sequence is evenly distributed between zero and a maximum. This maximum is called a *Contention Window* (CW) (where it might be better to replace the word *contention* with "competition" in this context. For more details, see section 4.3.5). The random number is multiplied by the time slot duration to define the duration of the backoff time (backoff counter). The time slot's duration depends on which physical transfer protocol is being used. Tables 4.5 and 4.7 list the base values. A special feature is that the remaining waiting time is not reduced if the medium remains busy. When the medium becomes free, the communications participant waits for a DIFS time interval and then periodically reduces the backoff time counter. The full CW value is reduced to zero once a transmission has been completed successfully.

This backoff algorithm is designed to prevent more than one other waiting station starting its transfer immediately after a station has finished its own transfer. However, as Figure 4.7 shows, collisions can still happen:

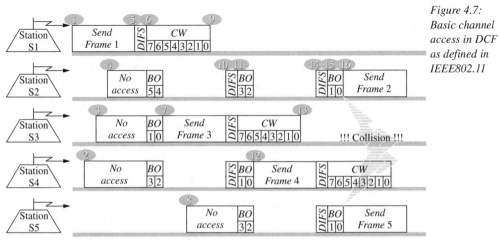

*Figure 4.7: Basic channel access in DCF as defined in IEEE802.11*

Situation without taking into account runtimes and without confirmation from the receivers (ACK)

**Notes on Figure 4.7**

1. Station S1 starts transmission of the first frame, *Frame 2*.

2. Physically remote Station S4 later detects that the channel is busy and delays its own access.

3. Physically remote Station S3 later detects that the channel is busy and delays its own access.

4. Physically remote Station S2 later detects that the channel is busy and delays its own access.

5. After Frame 1 has been transferred successfully all stations wait for a DIFS time period.

6. Station S1 starts a *Contention Window*. All the other stations, which want to access the channel, enter *Backoff* mode. The initial value of the backoff counter is selected as a random number within the time interval.

7. Station S3's backoff counter is the first to reach 0 and starts transmitting its frame *Frame 3*.

8. Physically remote Station S5 later detects that the channel is busy and delays its own access.

9. Station S1's *contention window* reaches 0. As no more data is ready to be transferred, this has no effect.

10. After *Frame 3* has been transferred successfully all active stations wait for a DIFS time interval.

11. Station S3 starts a *Contention Window*. All the other stations, which want to access the channel, enter *Backoff* mode.

12. Station S4's backoff counter is the first to reach 0 and starts transmitting its frame *Frame 4*.

13. Station S4's backoff counter is the first to reach 0 and starts transmitting its frame *Frame 4*.

14. Station S3's *contention window* reaches 0. As no more data is ready to be transferred, this has no effect.

15. After *Frame 4* has been transferred successfully all active stations wait for a DIFS time interval.

16. Station S4 starts a *Contention Window*. All the other stations, which want to access the channel, enter *Backoff* mode.

17. The backoff counters of Stations S2 and S5 reach 0 simultaneously and start transferring *Frame 2* and *Frame 5*. The electro-magnetic waves overlap and a collision occurs.

The value of CW is first set to an initial value $CW_{Min}$. After each unsuccessful transfer attempt it is then set to the next highest value in a pre-defined sequence. When a transmission is carried out successfully, CW is reset to $CW_{Min}$. When CW reaches its maximum value, $CW_{Max}$, it remains there until it is reset. CW is calculated by the formula $2^n-1$ in which the initial value $n$ depends on which physical transfer protocol is being used (see also Figure 4.8).

| Transmission attempt | 1 | 2 | 3 | 4 | 5 | 6 | 7 |
|---|---|---|---|---|---|---|---|
| n | 3 | 4 | 5 | 6 | 7 | 8 | 8 |
| CW | 7 | 15 | 31 | 63 | 127 | 255 | 255 |

CWmin                                          CWmax

*Figure 4.8: Selecting the length of the Contention Window (CW) depending on the number of previous unsuccessful transfer attempts*

## Acknowledgement mechanism

It is obvious that this algorithm is designed to suit the properties of wireless transmission:

- Firstly, collisions can still occur which destroy the data packets from both transmitting communications participants, as shown in Figure 4.9. (*Carrier Sense Multiple Access - Collision detection*) algorithm like the one present in Ethernet 8.2.3 cannot be used here.

- Secondly, wireless transfer systems do not have a method for detecting collisions. As a result a CSMA/CD (*Carrier Sense Multiple Access - Collision detection*) algorithm like the one present in Ethernet 8.2.3 cannot be used here.

For this reason, in wireless transmission, as defined in 802.11, the successful reception of a frame is acknowledged. The *acknowledgement* (ACK) is sent after a waiting time interval that is known as a *Short Interframe Space* (SIFS).

The only remaining case now is that Station S1 does not receive the acknowledgement within the pre-defined time interval. This can have (at least) two causes:

- The two data packets may perhaps have collided, as shown in Figure 4.7.

- It is also quite possible that the ACK frame has been destroyed. This may have happened when the two packets collided, which frequently happens if the total of the SIFS and the maximum spread time in the outbound and return direction is greater than the DIFS of a station that is ready to transmit. However, the packet may also have been disrupted by reflection (see also section 2.4.3).

If a station does not get back the acknowledgement frame at the right time, it prepares for retransmission by going into backoff state. During retransmission the *Retry* bit in the data frame's control field is used to signal that this transmission is a repeated attempt to send the data (see also section 4.3.6).

Figure 4.9 shows the process:

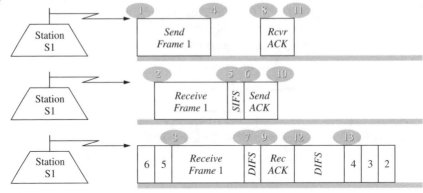

**Notes on Figure 4.9**

1. At a particular point in time Station 1 starts transmitting a frame, *Frame 1*, which is addressed to Station S2. At this time Station S3 in backoff mode.

2. Station S2 starts the normal reception of *Frame 2*.

3. Station S3 also receives the frame. Once the medium is recognised as busy the backoff counter is no longer decremented.

4. Station S1 ends transferring its frame.

5. Station S2 detects that the medium is free again. As the transmission ended properly, an SIFS wait interval starts so that Station S2 can send an acknowledgement.

6. Once the SIFS has ended Station S2 sends the ACK frame.

7. Station S3 detects that the medium is free again and starts a DIFS wait interval.

8. Station S1 receives the ACK frame.

9. Station S3 receives the ACK frame and realises that the medium is busy. It interrupts the DIFS wait interval which has still not run until its end.

10. Station S2 stops transferring the ACK frame.

11. Station S1 stops receiving the ACK frame.

12. Station S1 stops receiving the ACK frame and starts a new DIFS wait interval.

13. Once this has ended Station S1 can decrement the backoff counter again.

**Interframe Spaces**

In the access mechanism functionality the time between two frames, the *Interframe Space* (IFS), plays a central role. To determine if the medium is busy a communications participant must monitor the traffic on the medium during the IFS time period. In the standard, four different IFS times are defined, involving three different priority levels for channel access. As a general rule: the shorter the IFS time, the higher its priority.

As a result, SIFS is shorter than DIFS so that the acknowledgement can be sent before the waiting times of normal data transfer. As the SIFS waiting time is shorter, the acknowledgement frames are assigned higher priority than normal data packets, which must wait for the longer DIFS time.

The four types of IFS are shown below, in Table 4.1.

|  |  | **Function** | **Duration** |
|---|---|---|---|
| SIFS | Short IFS | ACK frame, CTS frame, PCF polling | Depending on which PHY is used |
| PIFS | PCF IFS | If, at the start of the CFP, a station in the PCF wants priority access to the medium | SIFS + SlotTime |
| DIFS | DCF IFS | Wait time in DCF frame | SIFS + 2 * SlotTime |
| EIFS | Extended IFS | Various functions | SIFS + DIFS + ACK control frame |

*Table 4.1: The different lengths of the Interframe Spaces (IFS) set the priority for channel access*

## 4.3.3    RTS-CTS mechanism

The procedure described in the sections above is reliable as long as every station is in radio contact with all other stations. If they are not, however, it can happen that one of the stations recognises the medium as free, even though this does not apply for another station. Figure 4.10 shows a possible scenario, including the presence of "hidden stations" of this kind. Station S1 can receive data from two other stations, S2 and S3. However, direct radio contact between Station S2 and Station S3 is impossible. If Station S3 sends a message to Station S2, the medium appears to be free, to Station S2. If S1 were now to transfer a data packet to S2 then S2 would not be able to receive it without errors.

*Figure 4.10:*
*Arrangement of*
*hidden*
*stations.*
*The radio cells*
*of stations S3*
*and S4 do not*
*overlap*

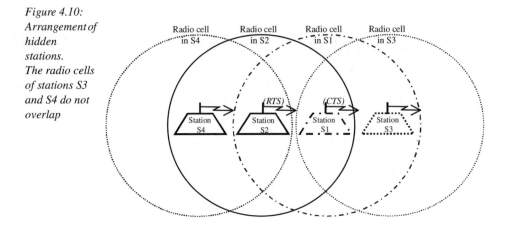

To avoid such situations the 802.11 protocol contains the "RTS-CTS" mechanism. The station that wants to transmit sends a ***Request-To-Send*** frame (RTS) to the recipient. The recipient responds by sending back a ***Clear-To-Send*** frame (CTS). If the transmitter successfully receives the CTS frame, the data is transmitted after a SIFS time period. If the CTS frame is not received within the defined time interval, the RTS frame must be sent again after the execution of a normal *Backoff* cycle. Using the SIFS time guarantees that the CTS reply has a higher priority than normal data transmission.

The RTS-CTS mechanism also involves two other facts:

*   Implementation of the RTS mechanism is optional. Nevertheless each device that complies with 802.11 must be able to reply to a RTS frame within the required time with the corresponding CTS frame.

*   RTS and CTS frames contain a field that shows how long it should take for the data frame to be transferred. This information is evaluated by any station that can receive RTS or CTS frames. In these stations a *Net Allocation Vector* (NAV) time counter is activated. Unlike the backoff counter this counter decrements independently of the transfer medium's state. During the NAV time none of these stations start a transfer procedure. As a result the likelihood of collisions is significantly reduced. This logical blocking of channel access is also called *virtual sensing* because the NAV counter is monitored at the same time.

    Note, however, that the NAV counter is not only set in the RTS-CTS mechanism, but by the *length* field in each individual frame (see also section 4.3.6).

In Figure 4.10, Station S4 can evaluate the RTS frame and Station S3 can evaluate the CTS frame. Figure 4.11 shows the corresponding timeflow without taking the medium's signal runtimes into account.

Clearly, this mechanism can therefore provide protection against collisions after the RTS and CTS packets have been successfully transferred. However, the price for this protection is an increase in protocol traffic on the channel. As this increase is especially disruptive for short data packets, a maximum length has been set to optimise this procedure. The RTS-CTS mechanism is only used to transfer data packets that exceed this size (*RTS threshold*).

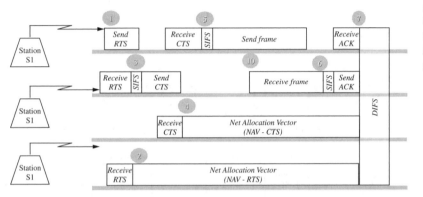

*Figure 4.11:
Channel access
when using the
RTS-CTS
mechanism*

**Notes on Figure 4.11**

1. At a particular point in time Station 1 starts transmitting an RTS frame that is addressed to Station S2. This frame can be received by stations S2 and S4.

2. After Station S4 successfully receives the RTS frame it sets the NAV time counter and "stays quiet". The NAV counter on Station S4 then gets the time it requires for the CTS, data and ACK frames in addition to the corresponding SIFS intervals.

3. Later on, after a SIFS time interval, Station S2 transmits the CTS frame. This can be received by Stations S1 and S3.

4. Station S3 sets its NAV counter to the time required to transfer the data and ACK frames, as well as the intervening SIFS time, and then "stays quiet".

5. After Station S1 has successfully received the CTS frame, and waited for a SIFS interval, it then transfers the data frame.

6. After waiting for a SIFS interval, Station S2 sends an ACK frame to confirm successful reception.

7. The data has been transferred successfully, the NAV counters in both stations that were not involved in the communication have run out and, after a DIFS time period has elapsed, a new data transmission can start.

The RTS-CTS mechanism has also proven useful when overlapping BSS and IBSS are used.

### 4.3.4    Fragmentation of long messages

Longer messages can be split into fragments to increase the likelihood of each individual fragment being transferred successfully.

In this situation each fragment that belongs to a message is marked with a sequence control field that is transferred as part of the control frame (see also section 4.3.6). In some cases the message is given an identification number, and in others each frame is given a sequential number. In addition, the *More Frag* bit is set for each frame. Only the very last frame does not have this identification. Figure 4.12 provides an overview of this process.

*Figure 4.12: Fragment- ation of long messages*

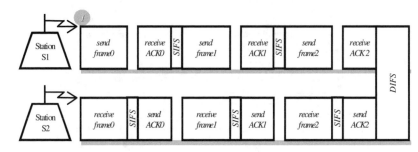

If a message is not a broadcast message, the receiving station returns an acknowledgement. After the DIFS time period has elapsed, and before other stations can access the channel, the transmitting station must send the next fragment immediately after a SIFS interval. This guarantees that the transmitting station has continuous access for subsequent time slots.

If the transmission is interrupted either by the data packet or the acknowledgement, the transmitting station does not receive an acknowledgement. The transmitting station then tries the transfer again (like in the base protocol), but this time the *retry* bit is set in the control frame (see also section 4.3.6).

Channels can also be reserved for fragmented messages (see also Figure 4.13). In this case the RTS packet from the transmitting station details how long the first fragment *frame 0* should take to transmit. In *frame 0* from the transmitting station the Duration/ID field in the header transfers the extension of the NAV vector for the next fragment. In the same way, the Duration/ID field contains the acknowledgement for this fragment so that all the stations in the reception zone can delay their transmissions for the next fragment as well. The data and acknowledgement packets can be regarded as virtual RTS/CTS packets. In the last fragment, *frame 2*, the reservation only still contains the time for a SIFS and an acknowledgement.

The system administrator can set the *RTSthreshold* value, above which a packet should be fragmented. Two aspects must be considered here. Firstly, long packets are prone to bit errors during transmission and, if errors occur, the entire transmission must be repeated. Secondly, fragmentation increases the burden on the protocol and therefore tests are needed to find the best possible setting to suit the particular features of each radio network. It is usually necessary to experiment with different settings, and see which are most suitable.

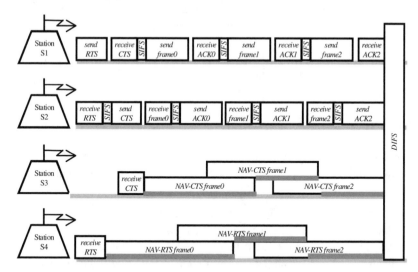

*Figure 4.13:*
*Fragmentation*
*of long*
*messages using*
*the RTS-CTS*
*algorithm*

## 4.3.5    Point Coordination Function

PCF was defined to support time-critical services. It provides a *Point Coordinator* (PC) with prioritised access to the transfer medium. Usually an AP with a fixed network access operates as a PC.

Although PCF is optional, all stations must be capable of using the PCF sequence media access rules, firstly because these are based on DCF, and secondly because, if they did not, this principle would have no effect. Stations that can reply to a PC's queries are called *CF pollable* stations. Besides the AP, only these stations can transfer data in accordance with PCF. However, you should be aware that PCF has not been implemented in most currently-available devices in an attempt to reduce costs.

PCF controls a frame's transfer during a *Contention Free Period* (CFP), which alternates with DCF-controlled *Contention Periods* (CP). The CFP is repeated at regular intervals at the CFP rate and starts with the transmission of a beacon frame (see also section 4.3.6) which contains the CFP's maximum duration. All stations in the BSS (apart from the PC) set their NAV to this value, as illustrated in Figure 4.14, on the next page.

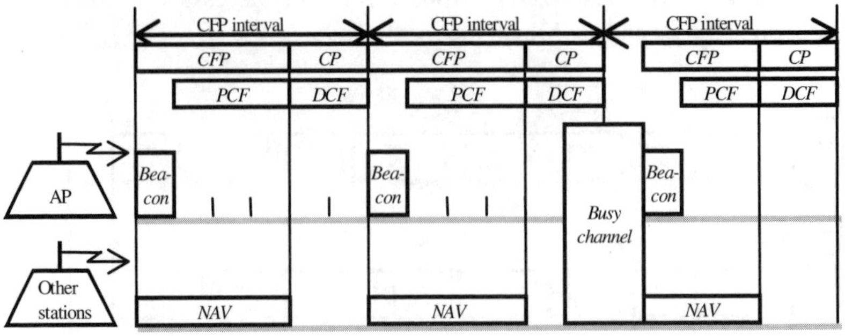

*Figure 4.14:*
*Reserving a*
*channel using*
*the Point*
*Coordination*
*Function*

After PCF IFS (PIFS) the PC takes control at the start of the CFP, once the transfer medium has been identified as free. PIFS is longer than SIFS but shorter than DIFS.

This means that PCF has a higher priority than normal transmission and must wait for an acknowledgement after the SIFS has been transferred. The PC remains in control throughout the entire CFP. It manages what is known as a "polling list" and, during the CFP, queries all the stations on this list in turn to find out whether they require a transfer. This centralised assignment of access rights to the transmission channel guarantees that no collisions can occur. In this scenario the PC uses an *association identifier* (AID) to identify each station. When a station logs on at the access point it can specify that its details should be included in the polling list.

Figure 4.14. illustrates the possible flow of a CFP. In this flow, a PC can transmit the following packet types, in any combination, to the CF-pollable stations:

- Data to a station $S_x$. If no data is present for the station, this element can be omitted.
- CF call (CF poll) to a station $S_x$. If this station should not be called, this element can also be omitted.
- CF acknowledgement (CF-ACK) for the previous data transmission from Station $S_{x-1}$ to the PC. If Station $S_{x-1}$ has not transferred any data in the previous time slot, the PC does not send an acknowledgement.
- In addition, it can also send control frames. One of the possible control frames is the CF-End signal which signals that the CFP is complete and resets the NAV vector (counter) in the other stations involved. If the CFP is terminated in this way, before it has expired, this removes the channel reservation for the CP without any time delay. CF-END can be combined with CF-ACK.

Basically, the participating stations can transmit two types of packet after they have been called from the PC:

- Data to the PC.
- CF acknowledgements (CF-ACK) for the PC together with data contained in the call.

If a called station neither receives data from the PC nor wants to send data to the PC, it acknowledges the call with a null frame.

If a station fails to respond to a call, the PC continues the polling process with the next station, after the PCF-IFS time.

PCF seems complicated, especially in situations in which neighbouring access points working on the same channel have similar CFP intervals. To avoid the risk of regular, repeated, collisions, PC backoff times can be set by means of special random algorithms. The start of CFP can also be delayed if another transmission is not yet complete.

PCF offers a simple and fairly efficient basis for implementing protocols for time-critical services. Once again, however, it highlights the restriction of the 802.11 standard to the two lowest OSI layers, as the provision of quality of service occurs on the third layer.

A further limitation is that PCF has not yet been clearly defined in the standard. In particular, the regulations governing the processing of a polling list, and responses to a collision during CFP, have not yet been defined. In addition, because the length of CFP cannot always be adhered to, due to prior transmissions, this presents a serious limitation to the reliable provision of channel capacity for an application.

## 4.3.6    Frame formats in the MAC layer

**General frame formats**

Before the description of MAC-layer frame formats, below, here are some notes on the hierarchical organisation of a communication process. For devices that comply with IEEE802.11 this means that frame information is added in both the MAC and PHY layers (see also Figure 4.15). The MAC frames are discussed in this section. The PLCP frames in the PHY layer are introduced in sections 4.4.2 and 4.4.3.

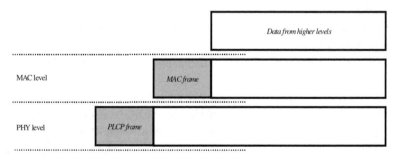

*Figure 4.15: Hierarchical frame structure in IEEE802*

Three different frame types are used to organise channel access: data frames, control frames and management frames. Figure 4.16 (1) illustrates the general format of a frame at MAC level, which corresponds to a data frame format.

*Figure 4.16:*
*(1) Format of a*
*general MAC*
*frame.*
*(2) Format of a*
*control field in*
*a data frame.*
*(3)Possible*
*frame types*
*and their*
*encoding*

(1)

| | | | MAC header | | | | | | |
|---|---|---|---|---|---|---|---|---|---|
| Control | Length | Address 1 | Address 2 | Address 3 | Sequence control | Address 4 | Data | Check-sum |
| 2 Bytes | 2 Bytes | 6 Bytes | 6 Bytes | 2 Bytes | 2 Bytes | 6 Bytes | 0 to 2312 Bytes | 4 Bytes |

(2)

| Version | Type | Subtype | To DS | From DS | More Frag | Retry | Power Mgmt | More Data | WEP | Order |
|---|---|---|---|---|---|---|---|---|---|---|
| 2 bits | 2 bits | 4 bits | 1 bits | 1 bit | 1 bit | 1 bit | 1 bit | 1 bit | 1 bit | 1 bit |

(3)

| 00 | 01 | 10 | 11 |
|---|---|---|---|
| Management frame | Control frame | Data frame | |
| to manage the cells | to regulate access to the medium and to ensure reliable transfer (e.g. ACK frames) | to transfer useful data | reserved |

In this context, please note that

- *Length:* specifies a frame's length in order to reserve the channel by using the NAV vectors.
- *Sequence:* is responsible for numbering fragmented messages.

Here the control field in a data frame is structured as shown in Figure 4.16 (2). In this case the individual fields have the following meanings:

- *Version:* shows the version of the IEEE802.11 protocol in use. At present only the value "00" is permitted here.
- *Type:* uses the coding from Figure 4.16 (3) to show the frame's type.
- *Subtype:* describes the frame's meaning, based on the coding from Figure 4.17, depending on the type of frame.
- *To DS, From DS:* describes the transition of data frames between a cell and the *Distribution System* (DS) in the context of the address fields, as shown in Figure 4.18. Here the cell address corresponds to the MAC address on the radio side of the access point
- *More Frag:* is set to "1" if more frames that belong to the same message are to follow in fragmented messages. If the current frame is the final frame of a fragmented message, or the only frame of a non-fragmented message, this bit has the value "0".
- *Retry:* is set to "1" when a frame is sent again.
- *Pwr Mgmt, More Data:* used to manage power-saving modes.
- *WEP:* if the WEP bit is set to "1" the data part of the frame is transmitted in code using the Wired Equivalent Privacy algorithm (see section 4.6).

- *Order:* this bit specifies that a frame is to be transferred to the upper layers in the correct sequence of messages (optional functionality).

| Sub-type value | Data frames | Control frame | Management frame |
|---|---|---|---|
| 0000 | Data | | Association Request |
| 0001 | Data + CF-ACK | | Association Response |
| 0010 | Data + CF-Poll | | Reassociation Request |
| 0011 | Data + CF-Ack + CF-Poll | | Reassociation Response |
| 0100 | Null | | Probe Request |
| 0101 | CF-Ack | | Probe Response |
| 0110 | CF-Poll | | |
| 0111 | CF-Ack + CF-Poll | | |
| 1000 | | | Beacon |
| 1001 | | | ATIM (Power Save) |
| 1010 | | Power Save (PS) Poll | De-association |
| 1011 | | RTS (Request-To-Send) | Authentication |
| 1100 | | CTS (Clear-To-Send) | De-authentication |
| 1101 | | ACK (Acknowledgement) | |
| 1110 | | CF-End | |
| 1111 | | CF-End + CF-Ack | |

Figure 4.17: Possible subtypes and their encoding

| To DS | 0 | 0 | 1 | 1 |
|---|---|---|---|---|
| From DS | 0 | 1 | 0 | 1 |
| Address 1 | Receiver | Receiver | Transmitter's cell | Receiver's cell |
| Address 2 | Transmitter | Receiver's cell | Transmitter | Transmitter's cell |
| Address 3 | Cell | Transmitter | Receiver | Receiver |
| Address 4 | - | - | - | Transmitter |
| Meaning | The frame is only sent within one cell. This applies to all control and management frames. | The frame is sent from the distribution system to a station in the cell via the access point. | The frame is sent by a station from a cell to the access point so that it can be passed on into the distribution system | The frame is sent by a station in a cell via its own access point to a station in another cell. |

Figure 4.18: Possible transitions between cells and the distribution system and their coding

### Data frames

The format of data frames follows the general details given above.

### Control frames

The subtypes for control frames are also shown in Figure 4.17. The frame format of control frames depends on the particular subtype. However, the control field in the header has the same structure in all control frames (see also Figure 4.19).

Control frames are only transmitted within a cell and are not encrypted.

*Figure 4.19:*
*Format of the*
*control field in*
*control frames*

| Version | Control frame | Subtype | To DS | From DS | More Frag. | Retry | Power Mgmt. | More Data | WEP | Order |
|---------|---------------|---------|-------|---------|------------|-------|-------------|-----------|------|-------|
| Version | Type | Subtype | 0 | 0 | 0 | 0 | Pwr Mgmt | 0 | 0 | 0 |
| 2 bits | 2 bits | 4 bits | 1 bit | 1 bit | 1 bit | 1 bit | 1 bit | 1 bit | 1 bit | 1 bit |

## Management frames

The management frame format shown in Figure 4.20, below, is independent of the subtype shown in Figure 4.17. In particular, *Beacon* frames are a kind of management frame.

*Figure 4.20:*
*Format of*
*management*
*frames*

| MAC header | | | | | | Data | Checksum |
|------------|--------|----------|-------------|------|------------------|----------------|----------|
| Control | Length | Receiver | Transmitter | Cell | Sequence control | | |
| 2 Bytes | 2 Bytes | 6 Bytes | 6 Bytes | 6 Bytes | 2 Bytes | 0 to 2312 Bytes | 4 Bytes |

Finally, please note that the control information described here

- only represents part of the complete configurations described in the standard.
- does not include numerous other proprietary extensions, which in many cases have not been made public.

## 4.3.7   Managing stations

This section is devoted to two topics in particular:

- The assignment of mobile stations to a cell in an access point. Here, please note that the meaning of the term "cell", as used in the IEEE802.11 standard, is used to describe a channel cell as defined in section 2.4.3.
- The structure of ad hoc networks.

Managing stations in the context of ad hoc networking, as described in section 2.3.10, involves the following three steps:

- station identification
- authentication
- logon.

The aspects of managing stations that involve addressing and identifying them are described in section 4.3.8.

## Beacon frames

Beacon frames are management frames (see also section 4.3.6), in which a cell is sent at periodic time intervals per broadcast to all the users in a cell and contains basic information about the cell itself. This is illustrated in Figure 4.21.

| Information | Length | Description | | |
|---|---|---|---|---|
| Timestamp | 8 Bytes | Transmitter time used to synchronize the clocks in all the stations in one cell | | |
| Beacon interval | 2 Bytes | The time period between two beacons. Together with the time, each station can calculate the start time of the next beacon. This is called the "Target Beacon Transmission Time" (TBTT). As a result,a station can also recognise that a beacon is not present. | | |
| Capability | 8 Bytes | Information about the cell: ad hoc or infrastructure, CFP-supported, encryption | | |
| Address | 6 Bytes | Cell address: in an infrastructure network the cell address corresponds to the MAC address of the access point | | |
| Data rate | 3 to 8 Bytes | | | |
| FH parameter | 7 Bytes | DS parameter | 7 Bytes | Alternative transmission of PHY layer properties: channel selection |
| CF Parameter | 8 Bytes | If the access point implements PCF, data such as the starttime of the next CFP, the maximum duration of the CFP or the remaining time of the current CFP is also transferred. | | |
| TIM | 4 Bytes | Transmission Indication Map for controlling power-saving mode | | |

*Figure 4.21: Format of beacon frames*

Beacon frames are sent by access points in infrastructure networks. In ad hoc networks each station can send a beacon frame every TBTT (*Target Beacon Transmission Time*).

To prevent several stations from sending a beacon frame simultaneously an algorithm is used, once again, for collision avoidance: each station loads a timer containing a random number, on the next TBTT, and only sends the beacon frame when the timer has expired, and providing that the station has not received a beacon frame from another station in the meantime.

If an access point is switched on it begins to send beacon frames at fixed regular intervals. A station that is not an AP can therefore monitor all physical channels to find out from which cells it can receive signals. This process is known as *passive scanning*. As a result one station can receive beacon frames from several access points, if it is located in an area in which cells overlap. Using the beacon's signal strength the station can decide which cell it wants to log on to. If the station is configured so that it can only log on to a particular cell, this restricts the selection it can make.

However, a station can also be configured in such a way that it starts a new cell as an ad hoc network, after it has waited for a certain period of time (which can be changed as required) to receive a beacon frame for an ad hoc network (see also section 9.2.3).

In addition, stations can also use *active scanning*. In this case the management frames in each broadcast are sent with the *Probe Request* subtype. The data part of the frame either states that the station is searching for any cell, or contains the address of a specific cell that the station is looking for. When an AP receives this type of frame it replies with a management frame, with the *Probe Response* subtype, containing information about its cell. In an ad hoc network the station that responds is the one that has most recently generated a beacon frame for the cell.

## Authentication

Authentication is the way that one station proves its identity to another. The IEEE802.11 standard specifies two types of authentication:

- open authentication. This uses a very simple algorithm, which only fulfils the requirements of authentication in a very superficial way. The open authentication shown in Figure 4.22 is the most commonly-used procedure:

*Figure 4.22: Open authentication by stations*

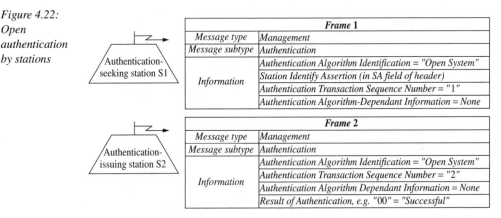

- authentication using a common key (*shared key authentication*). This is based on encrypted communications using the WEP algorithm. It involves a check to see whether both participating stations have the same secret key.

In an open authentication check, Station S2 (the one that is issuing the authentication) transfers a text to Station S2 (the one looking for authentication). This is a *challenge text* which is then encoded and returned by Station S2. This message is only authenticated if the encryption is correct (see also Figure 4.23).

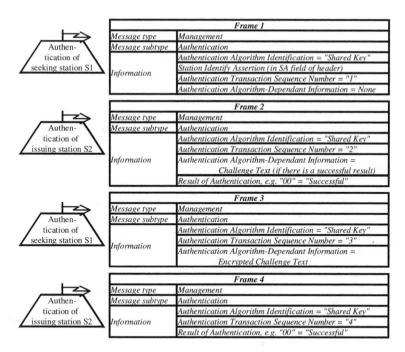

## Logon

The crucial factor in ensuring that infrastructure networks function correctly is the assignment of the stations to the appropriate cells. This means that each station must be certain which cell it belongs to, in order to send data to the distribution system. ESSID or SSID can be used to make this kind of assignment. However, in many cases manufacturer-specific control information is also used.

On the other hand, an access point must know which stations it is to recognise as belonging to its cell. This assignment is carried out by an explicit logon procedure (*association*), as shown in Figure 4.24:

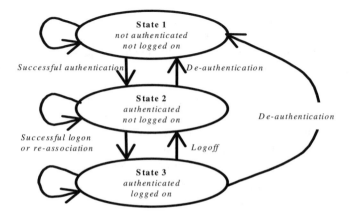

*Figure 4.24:*
*Algorithm of an*
*explicit logon*
*(association)*
*procedure*

Together, the logon and authentication stages form a two-level assignment system:

- A station can only log on if it has been authenticated.

- A station can only use the distribution system after it has logged into a cell.

The IEEE802.11 standard also describes the classification shown in Figure 4.25. This illustrates which frame can be sent or received in each of the three station states to implement this two-level structure. The thing to note here is that data frames can be sent and received within a cell even in state 1 (not authenticated and not logged on).

*Figure 4.25: Classification of rights depending on the logon status*

| Class | Data frame | Control frame | Management frame | State 1 | State 2 | State 3 |
|---|---|---|---|---|---|---|
| 1 | all within the cell (ToDS = FromDS = 0) | all except PS poll | all except • Association Request/Response • Re-Association Request/Response • Disassociation | permitted | permitted | permitted |
| 2 | – | – | all the rest, i.e • Association Request/Response • Re-Association Request/Response • Disassociation | not permitted | permitted | permitted |
| 3 | all data frames in which the ToDS or FromDS bits are set | PS Poll | – | not permitted | not permitted | permitted |

## 4.3.8 Addressing stations

The addressing of frames as defined in IEEE802.11 networks differs in two ways from addressing in the MAC layer and identification at network level. Despite this, the procedures are related to each other.

In the MAC layer IEEE802.11 uses addresses which correspond to the MAC addresses defined in the IEEE802.3 Ethernet standard. For this reason, integration into existing operating systems is relatively simple. However, there is another aspect that should be mentioned here because it sometimes causes confusion. Stations in radio networks that act as bridges have at least two network connections. Access points are an example of this. These stations are, on the one hand, connected to the wired *Distribution System*, on which they have a MAC address for the IEEE802.3 network. On the other hand they also have a radio module to which a MAC address is also assigned. So, the access point addresses listed in Figure 4.17 are in fact radio module addresses.

An identification procedure is also carried out at network level. There are two main identification levels used here:

- Each *Basic Service Set* (BSS) has a unique BSSID assigned to it to identify it. In a wired BSS the access point's MAC address is selected as it provides the service for the BSS. For an *Independent BSS* (IBSS) the BSSID is generated at random.

- This SSID is also an element used to restrict access (see also section 4.6).

# 4.4    The Physical layer and bit transfer

## 4.4.1    Where it fits in

As you might expect, the greatest differences between wired and wireless communications are to be found in the Physical layer. This is because electro-magnetic waves are used as the transfer medium through the "air interface". When specifying the physical transfer protocol, it is vital that the properties of transmission via the air interface (especially noise and other interferences) are taken into account.

Whereas earlier IEEE802.11-compatible products mainly used the FHSS procedure, the DSSS procedure is becoming ever more popular nowadays. The primary reason for this is that extended IEEE802.11b and its data rate of 11 Mbps can only be achieved with DSSS. This provides a simple way of ensuring that systems stay backwards-compatible with slower transfer speeds.

## 4.4.2    Frequency Hopping Spread Spectrum

The FHSS standard, as defined in IEEE802.11, includes up to 79 non-overlapping frequency ranges, each with a bandwidth of 1 MHz. It contains three groups, each with 26 patterns. The sequence of frequencies is calculated from a base sequence that corresponds to a pseudo random chain in the interval from 0 to 78.

For example, base sequence b(i) is defined for North America and Europe in the values listed in Table 4.2:

| i | 1 | 2 | 3 | 4 | 5 | 6 | 7 | 8 | 9 | 10 | 11 | 12 | 13 | 14 | 15 | 16 | 17 | 18 | 19 | 20 |
|---|---|---|---|---|---|---|---|---|---|----|----|----|----|----|----|----|----|----|----|----|
| b(i) | 0 | 23 | 62 | 8 | 43 | 16 | 71 | 47 | 19 | 61 | 76 | 29 | 59 | 22 | 52 | 63 | 26 | 77 | 31 | 2 |
| i | 21 | 22 | 23 | 24 | 25 | 26 | 27 | 28 | 29 | 30 | 31 | 32 | 33 | 34 | 35 | 36 | 37 | 38 | 39 | 40 |
| b(i) | 18 | 11 | 36 | 71 | 54 | 69 | 21 | 3 | 37 | 10 | 34 | 66 | 7 | 68 | 75 | 4 | 60 | 27 | 12 | 25 |
| i | 41 | 42 | 43 | 44 | 45 | 46 | 47 | 48 | 49 | 50 | 51 | 52 | 53 | 54 | 55 | 56 | 57 | 58 | 59 | 60 |
| b(i) | 14 | 57 | 41 | 74 | 32 | 70 | 9 | 58 | 78 | 45 | 20 | 73 | 64 | 39 | 13 | 33 | 65 | 50 | 56 | 42 |
| i | 61 | 62 | 63 | 64 | 65 | 66 | 67 | 68 | 69 | 70 | 71 | 72 | 73 | 74 | 75 | 76 | 77 | 78 | 79 | - |
| b(i) | 48 | 15 | 5 | 17 | 6 | 67 | 49 | 40 | 1 | 28 | 55 | 35 | 53 | 24 | 44 | 51 | 38 | 30 | 46 | - |

Table 4.2: Base sequences for FHSS in accordance with IEEE802.11 in North America and Europe

The transfer frequency of the pattern derived from this base sequence using the calculation rules in Table 4.3 is expressed as:

$$F_x(i) = 2400 + f_x(i) \text{ [GHz]}$$

Table 4.3:
Assignment of
frequency
hopping
sequences

| Region | Released frequency band | Hopping frequencies | Channels | Channel number calculation |
|---|---|---|---|---|
| USA | 2.4000 to 2.48365 Ghz | 2.402 to 2.480 Ghz | 2 to 80 | [b(i) + x] mod 79 + 2 |
| Europe | 2.4000 to 2.48365 Ghz | 2.402 to 2.480 Ghz | 2 to 80 | [b(i) + x] mod 79 + 2 |
| Japan | 2.4710 to 2.4970 Ghz | 2.473 to 2.495 Ghz | 73 to 95 | [(i-1) * x] mod 23 + 73 |
| Spain | 2.4465 to 2.4835 Ghz | 2.447 to 2.473 Ghz | 47 to 73 | [b(i) + x] mod 27 + 47 |
| France | 2.4450 to 2.4750 Ghz | 2.448 to 2.482 Ghz | 48 to 82 | [b(i) + x] mod 35 + 48 |

The offset $x$ of the channel pattern is required here. For North America and Europe this is divided into the three groups shown in Table 4.4:

Table 4.4:
Assugnment of
frequency
hopping
sequences

| Set 1 | 0 | 3 | 6 | 9 | 12 | 15 | 18 | 21 | 24 | 27 | 30 | 33 | 36 | 39 | 42 | 45 | 48 | 51 | 54 | 57 | 60 | 63 | 66 | 69 | 72 | 75 |
|---|---|---|---|---|---|---|---|---|---|---|---|---|---|---|---|---|---|---|---|---|---|---|---|---|---|---|
| Set 2 | 1 | 4 | 7 | 10 | 13 | 16 | 19 | 22 | 25 | 28 | 31 | 34 | 37 | 40 | 43 | 46 | 49 | 52 | 55 | 58 | 61 | 64 | 67 | 70 | 73 | 76 |
| Set 3 | 2 | 5 | 8 | 11 | 14 | 17 | 20 | 23 | 26 | 29 | 32 | 35 | 38 | 41 | 44 | 47 | 50 | 53 | 56 | 59 | 62 | 65 | 68 | 71 | 74 | 77 |

In regions where the ISM bandwidth is narrower there are fewer frequency ranges (see also Table 4.3), and as a result the possible density of devices is also lower. As a consequence a smaller number of patterns are available for calculating the base sequences and channel patterns.

Figure 4.26:
Frame format
of the Physical
Layer
Convergence
Procedure
frame (PCLP)
for the FHSS
procedure as
defined in
IEEE802.11

| PPDU | | | | | | |
|---|---|---|---|---|---|---|
| PLCP Protocol Data Unit | | | | | | |
| PLCP Preamble | | PLCP Header | | | | MPDU |
| Synchronisation | Start Frame Delimiter | MPDU Length Word | PLCP Signaling Field | Header Error Check | | MAC Protocol Data Unit |
| SYNC | SFD | PLW | PSF | HEC | | 0 to 4095 Bytes |
| 80 bits | 16 bits | 12 bits | 4 bits | 16 bits | | |
| Consists of alternating zeros and ones; is used by the receiver to recognise an incoming signal and for synchronization | Fixed bit pattern that marks the start of the actual frame | Specifies the MPDU length in bytes; maximum length: 212-1 = 4095 Bytes | Specifies the data rate at which the MPDU is sent | Error check for the PLCP header using 16b CRC | | Useful data |

The frame format of the FHSS transmission route is shown in Figure 4.26. Here the preamble and header are usually transferred at a data rate of 1 Mbps whereas the data packets can be sent at 1 or 2 Mbps. The operating type that uses 1 Mbps is defined for all devices in the original 802.11 standard of 1997. Transfer at 2 Mbps is a later development and is optional. The increase in data rate was achieved by *Multilevel Signalling* within GFSK (*Gaussian Phase Shift Keying*) modulation in which the symbol speed remains constant at 1 MBaud.

During a 1 Mbps transfer one bit is transferred per symbol, but during a 2 Mbps transfer two bits are grouped in one symbol (see Figure 4.27).

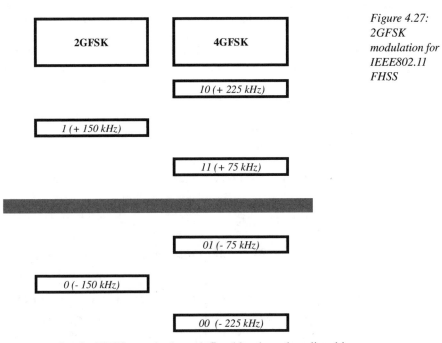

*Figure 4.27: 2GFSK modulation for IEEE802.11 FHSS*

The basic parameters for the FHSS standard are defined by the values listed in Table 4.5.

| Size | Value |
|------|-------|
| AslotTime | 50 s |
| ASIFSTime | 28 s |
| AMPDUMaxLength | 4095 bytes |

*Table 4.5: Basic FHSS transmission parameters in IEEE802.11 standard*

## 4.4.3   Direct Sequence Spread Spectrum

In the DSSS procedure, frequency spread is achieved by a logical link with a pseudo-noise (PN) sequence that has a higher bit rate. In DSSS systems that comply with the 802.11 standard the 11-chip long Barker Code is used for data rates of 1 and 2 Mbps. This has particularly good auto correlation properties, and its length corresponds to the minimum length of a spread code as defined in the regulators' release guidelines.

The DSSS procedure as defined in IEEE802.11 uses the same code for all channels. The CDMA procedure is therefore only used to separate a transmission from other transmissions in the 2.4 GHz ISM band, and not to identify stations in the same radio cell that are being used simultaneously in other channels. The main reason for this is that the spread code length, and therefore the *Processing Gain* (GP), are too low to identify different codes. In this respect the 802.11 standard differs from other CDMA systems, as implemented in mobile communications, in which different PN codes can be used to separate the various channels. In many 802.11 systems different carrier frequencies are used to differentiate between the various channels, as part of an FDMA procedure. In this case a channel can achieve 22 MHz bandwidth when spreaded with an 11-bit PN code. Table 4.6 shows channel availability and other regional restrictions. In North America and in Europe this method can be used to operate three non-overlapping channels. Although only channels 1, 6 and 11, with an interval between bands of 3 MHz, shown in Figure 4.28, can be operated in a non-overlapping manner in North America, in Europe the 13 selection channels can provide different combinations or greater band intervals. For example, channels 1, 7 and 13 can be used to give a band interval of 8 MHz. If you use channels 1, 6 and 13, the band interval between channels 6 and 13 is still 30 MHz.

*Figure 4.28: Channel assignment in the DSSS procedure in IEEE802.11*

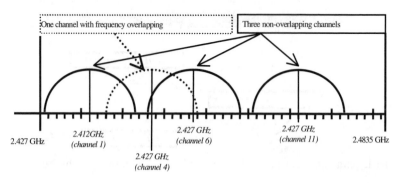

This aspect is often used by supporters of other frequency spread technologies to criticise the frequency hopping procedure. As section 2.7.3 explained, the quality of DSSS systems increases not just because of their channel efficiency, but also because of their sensitivity to disruption when the spread factor increases. Correspondingly the IEEE802.11 procedure only meets the minimum requirements.

The preamble and header are usually transferred at a data rate of 1 Mbps, although data packets can be set at 1 or 2 Mbps.

In addition, the data is arranged in a new sequence to make it possible to compensate for bundling errors more effectively. Unlike in voice transmission systems this *scrambling* is not used to provide a higher error tolerance in the case of bundling errors, but instead to equalise the frequency spectrum (*whitening*).

| Channel ID | Average frequency | North America | Europe | France | Spain | Japan |
|---|---|---|---|---|---|---|
| 1 | 2.412 GHz | x | x | | | |
| 2 | 2.417 GHz | x | x | | | |
| 3 | 2.422 GHz | x | x | | | |
| 4 | 2.427 GHz | x | x | | | |
| 5 | 2.432 GHz | x | X | | | |
| 6 | 2.427 GHz | x | X | | | |
| 7 | 2.442 GHz | x | x | | | |
| 8 | 2.447 GHz | x | x | | | |
| 9 | 2.452 GHz | x | x | | | |
| 10 | 2.457 GHz | x | x | x | x | |
| 11 | 2.462 GHz | x | x | x | x | |
| 12 | 2.467 GHz | | x | | x | |
| 13 | 2.472 GHz | | x | | x | |
| 14 | 2.484 GHz | | | | | x |

*Table 4.6: Channel assignment in the DSSS procedure in IEEE802.12. (Note: since the start of the year 2000 the entire spectrum from 2400 to 2483 MHz has also been available in France and Spain)*

Figure 4.29 shows the frame format for DSSS transfers.

| PPDU | | | | | | |
|---|---|---|---|---|---|---|
| PLCP Protocol Data Unit | | | | | | |
| PLCP Preamble | | PLCP Header | | | | MPDU |
| Synchronisation | Start Frame Delimiter | Signal | MPDU Length Word | PLCP Signaling Field | Header Error Check | MAC Protocol Data Unit |
| SYNC | SFD | PLW | PLW | PSF | HEC | 14 to 4095 Bytes |
| 128 bits | 16 bits | 12 bits | 12 bits | 4 bits | 16 bits | |
| Consists of alternating zeros and ones. Used by the receiver to recognise an incoming signal and for synchronization | Fixed bit pattern, that marks the start of the actual frame | Specifies the MPDU length in bytes; maximum length: 2¹²-1 = 4095 Bytes | Specifies the MPDU length in bytes; maximum length: 2¹²-1 = 4095 Bytes | Specifies the data rate at which the MPDU is sent | Error check for the PLCP header using 16b CRC | Useful data |

*Figure 4.29: Frame format of the Physical Layer Convergence Procedure (PLCP) frame for the DSSS procedure in IEEE802.11*

This higher bit-rate datastream is then modulated by a **Differential Quadrature Phase Shift Keying** (DQPSK) modulation, as shown in case a), a 1 Mbps transfer, in Figure 2.12. If the data rate is set to 2 Mbps the four states in case b) in Figure 2.12 are selected. The procedure described in section 4.7.2 is used to achieve even higher data transfer rates. The basic parameters for the DSSS standard are defined by the values listed in Table 4.7.

| Size | Value |
|---|---|
| AslotTime | 20 s |
| ASIFSTime | 10 s |
| AMPDUMaxLength | 4095 bytes |

*Table 4.7: Basic DSSS transmission parameters in IEEE802.11 standard*

## 4.5    Other services

### 4.5.1    Synchronisation

Synchronisation uses the *Timing Synchronisation Function* (TSF) to synchronise all stations in a network to the system time. To ensure synchronisation the TSF clock value is sent regularly, in a beacon, at the time intervals specified in the *Target Beacon Transmission Time* (TBTT). In an infrastructure network the access point is responsible for sending the synchronisation beacon, and in ad hoc networks all stations generate a beacon. In this way the beacon is transmitted by the different stations at varying, randomly-generated propagation delay times.

### 4.5.2    Power-saving mode

As many wireless devices are battery-operated a core element of the standard is a power-saving mode which a station must negotiate with other stations. In particular, stations must also be available for contact when in *Doze* mode. To achieve this certain monitoring algorithms have been developed. They are different for infrastructure and ad hoc mode.

### 4.5.3    ESS services

To operate *Extended Service Sets* (ESS) the following services must be provided:

*   the *distribution service*, used to transfer data between the transmitter and receiver access points
*   the *integration service*, used to transfer data from the transmitting station's access point to the receiver's portal in a fixed network, and modify the address and protocol if required, and
*   the *association service* which ensures that stations log on and off correctly when their BSS changes, and then informs all the access points in the distribution system about the new location.

## 4.6    Security

Security issues are particularly important for all wireless systems because the diffusion of electro-magnetic waves mean it is possible to listen or transmit at a physical level without actually entering a building without permission (*parking lot attack*). This is a major difference to wired transfer protocols, as in wired networks only external traffic can be monitored outside the building (on the other side of the switch or router).

On the other hand, the idea behind WLAN systems is to provide user-friendly *ad hoc networking* by implementing as much automatic support as possible for station identification, authentication and logon (authorisation).

In IEEE802.11-compliant WLAN systems there are three separate levels of security. However (as always) there are a few restrictions that cannot be ignored:

- At the lowest level users are given access by means of a code called the *Electronic System ID* (SSID, ESSID). The administrator sets this ID for all mobile users and all access points. This code shows a user's access rights but is not a unique identifier for that user. This results in two restrictions:
  - It is usually quite easy to work out a general access number in order to eavesdrop (illegally) on network traffic.
  - Most manufacturers of mobile stations allow the "*any*" option in their configuration files which authenticates the use of those mobile stations in any network.
- Authentication only permits authorised stations to participate in the communication process. In the 802.11 standard participating stations can exchange their identity with each other by using *Link Level Authentication*. To do so, the mobile users' MAC addresses are entered in the access lists that are held by the access points. However this has the following disadvantages:
  - In most currently-available products you can change the mobile user's MAC address, which opens a loophole for possible misuse.
  - Although authentication can be implemented quite easily in small wireless networks, larger networks with more access points will require time-consuming administration for each station, and for each user, so that users can move from one radio cell to the next (*roaming*). Until now only a very small number of manufacturers have provided easy-to-use tools for managing large-scale wireless networks.
- To support authentication for users, as well as at device level, also almost all manufacturers implement the **Remote Authentication Dial-In User Service** (RADIUS). This means user IDs and passwords can be managed centrally.
- As part of the optional **Wired Equivalent Privacy** (WEP) procedure, the contents of radio messages can be made confidential using 40-bit encryption. However you also need to be aware, that
  - WEP is only an optional part of the standard so you should ensure that it has been implemented in any 802.11-system that you might select.
  - Some manufacturers also offer 128-bit encryption, which is much better than normal 40-bit encryption. However, these are proprietary developments which do not help the cause of interoperability among systems from different manufacturers.
  - Even the tools used to manage the WEP key are manufacturer-specific.
  - The various crucial weaknesses of encryption using the WEP algorithm also mean that it is comparatively easy to decode. This is described in more detail in section 10.1.

The modifications that have been made to the IEEE802 standard also mean that all the security mechanisms used at higher protocol layers, such as IPSec, can also be implemented.

The WEP procedure is based on a symmetrical algorithm in which both the transmitter and receiver use a common key (the *shared key*). Starting with this key and a randomly-defined *Initialisation vector* (IV) a pseudo-random number generator is used to define a key sequence which is linked to the plain text bit-by-bit by an exclusive/or (XOR) association (see also Figure 4.30).

*Figure 4.30: Encryption, transmission and encryption using the WEP algorithm*

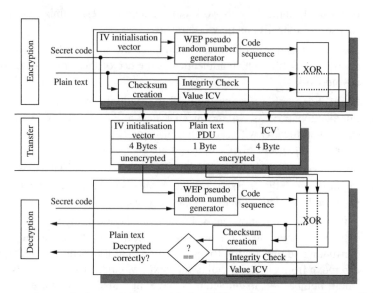

This generator works in accordance with the RC4 encryption algorithm, called the *Key Scheduling Algorithm* (KSA), which uses a 40-bit static key (for WEP) or 128-bit static key (for WEP2). This algorithm is based on the expansion of a comparatively short key into a pseudo-random key of infinite length which is then used to encrypt the text that is to be transmitted. In this case both the plain text and the checksum are encrypted and transferred together with the unencrypted initialisation vector. At the receiver end the encrypted text is decrypted, along with the expanded key, using an exclusive-or link.

This symmetrical algorithm is based on the fact illustrated in Figure A.2, that using a double exclusive-or relationship to link one signal sequence with a second signal sequence results in the original signal sequence. In Appendix A.2 this scenario is used for reception correlators in the direct sequence spread spectrum process.

Further security issues in the IEEE802.11 standard are discussed in section 10.1.

## 4.7   Extensions to the standard

### 4.7.1   The IEEE802.11a standard

The IEEE's 802.11a standard is an update of the original 802.11 standard. It has adopted the MAC-layer channel access mechanisms without making any changes to them. However, in the Physical layer, transmission now takes place in the 5 GHz range, and the rules governing the American *Unlicensed National*

*Information Infrastructure* (UNII) band played an important role in the definition of the new standard. Basically, data rates of 6 to 54 Mbps are planned. 802.11a is based on an *Orthogonal Frequency Division Multiplex* (OFDM) procedure (see also section 2.8), which in many respects corresponds to the definitions in the HiperLAN/2 standard and is therefore described in Chapter 8. Chapter 8 also includes a few comments about how these two technologies compete with each other.

The first devices to comply with 802.11a were announced at the end of 2002. However, due to their relatively high costs you can assume that the first applications will be seen in the LAN-LAN link area because the bandwidth requirement there is the highest and costs seem to be a less important factor.

As the channel access properties are identical, some manufacturers have already announced devices that are designed to facilitate migration to the more powerful systems [Proxim 2001].

## 4.7.2    The IEEE802.11b standard

The 802.11b standard, ratified in September 1999, and sometimes also known as 802.11HR (*High Rate*) in older sources, specifies systems with a bandwidth of 5.5 or 11 Mbps in the 2.4 GHz band. This has now become so popular that nowadays you can hardly find any systems that comply with the 802.11 standard and support data rates of only 1 or 2 Mbps on the market.

The IEEE802.11b standard is based on the DSSS procedure which makes these systems backwards-compatible with the 802.11-1997 DSSS standard. During the standardisation procedure there were serious conflicts of interest between supporters of FHSS in one camp and DSSS in the other, and a number of DSSS procedures are still under discussion. In an effort to protect their own investments the manufacturers Lucent and Harris (Prism) made the greatest attempts to influence the standardisation process in the 2.4 GHz range to use their products by developing proprietary systems. Despite this the standardisation committee did not select either Lucent's *Pulse Position Modulation* (PPM) with data rates of 5 and 10 Mbps or Harris's MBOK (*M-ary Bi-Orthogonal Keying*, where *m-ary=multiple*) with data rates of 5.5 and 11 Mbps (Heeg 1999).

The most important factor was that only one physical standard was specified to ensure the interoperability of all standard-compliant 11-Mbps devices. The principal means of increasing the data rate was to implement a modulation procedure that improved the way in which the frequency spectrum is used. This also means that a channel bandwidth of 22 MHz can still be used so that three independent channels can be operated. In particular, more bits per symbol can be transferred in the *Quadrature Phase Shift Keying* (QPSK) procedure than in *Binary Phase Shift Keying* (BPSK). In addition, shortened 8 bit chip sequences are used in QPSK.

In the case of higher data transfer rates the CDMA procedure is also used to distinguish between the transferred databits. The code words, which consist of eight complementary, complex chips, are calculated using common Hadamard encoding, in which the following is added:

- $\varphi_1$ to all chips
- $\varphi_2$ to all odd-numbered chips
- $\varphi_3$ to all odd-numbered pairs of chips and
- $\varphi_4$ to the first four chips.

The fourth and seventh chip are also given a minus sign prefix to improve the code's correlation properties.

$$\underline{C}=\left\{e^{j(\varphi_1+\varphi_2+\varphi_3+\varphi_4)},e^{j(\varphi_1+\varphi_3+\varphi_4)},e^{j(\varphi_1+\varphi_2+\varphi_4)},-e^{j(\varphi_1+\varphi_4)},\right.$$
$$\left. e^{j(\varphi_1+\varphi_2+\varphi_3)},e^{j(\varphi_1+\varphi_3)},-e^{j(\varphi_1+\varphi_2)},e^{j\varphi_1}\right\}$$

with $\varphi_i \in \{0,\pi/2,\pi,3\pi/2\}, i \in \{1,2,3,4\}$

In a 5.5 Mbps transfer each symbol consists of four databits. The first two bits $d_0$ and $d_1$ select phase shift $\varphi_1$, as shown in Figure 4.31, whereas $d_2$ and $d_3$ are used to specify the values of $\varphi_2$, $\varphi_3$ and $\varphi_4$ in accordance with

$$\varphi_2 = d_2 * \pi + \pi/2$$
$$\varphi_3 = 0$$
$$\varphi_4 = d_3 * \pi$$

When the transfer rate is doubled to 11 Mbps each symbol must contain eight bits. Once again, databits $d_0$ and $d_1$ define phase shift $\varphi_1$ as shown in Figure 4.31, whereas the pairs $d_2$-$d_3$, $d_4$-$d_5$ and $d_6$-$d_7$ set phases $\varphi_2$, $\varphi_3$ and $\varphi_4$ as shown in Figure 4.32, and therefore select one of the 64 possible CCK code words that is modulated by $\varphi_1$ DQPSK.

*Figure 4.31:*
*Setting DQPSK*
*modulation*
*phases at*
*5.5 Mbps*

| d0 | d1 | 1 | |
|----|----|---|---|
| | | *for the symbols c0, c2, c4 and c6* | *for the symbols c1, c3, c5 and c7* |
| 0 | 0 | 0 | ? |
| 0 | 1 | ?/2 | 3?/2 |
| 1 | 0 | ? | 0 |
| 1 | 1 | 3?/2 | ?/2 |

| $d_{J\pi}$ | $d_{J\pi}$ | |
|:--:|:--:|:--:|
| $d_{\pi}$ | $d_{J\pi}$ | $\varphi_i \cdot = \varphi_j \cdot = \varphi_o \cdot$ |
| $d_{o\pi}$ | $d_{J\pi}$ | |
| 0 | 0 | 0 |
| 0 | 1 | $\pi/2$ |
| 1 | 0 | $\pi$ |
| 1 | 1 | $3\pi/2$ |

*Figure 4.32: Setting DQPSK modulation phases at 11 Mbps*

This highlights the relationship between the symbol rate, which is increased to 2.375 MSymbols/s, and the reduced code length that results in an unchanged chip rate of 11 Mchips/s.

It is obvious that an increase in data transfer rate also makes the transfer more prone to disruption. The receiver must now be able to identify eight different states per symbol instead of two or four as was previously the case. In the same way, more demands are made on the signal-to-noise ratio (SNR): see also section 2.5. Table 4.8 shows typical values for the additional requirements placed on the SNR relative to the values for a successful transfer at 1 Mbps.

| | | Data rate | Chiprate |
|:--:|:--|:--:|:--:|
| 1 Mbps | 11 chips / symbol<br>1 bit / symbol<br>1 symbol / s | $1\dfrac{bit}{symbol}*1\dfrac{MSymbol}{s}=1\dfrac{Mbit}{s}$ | $11\dfrac{chips}{symbol}*1\dfrac{MSymbol}{s}=11\dfrac{MChips}{s}$ |
| 2 Mbps | 11 chips / bit<br>2 bits / symbol<br>1 symbol / s | $2\dfrac{bit}{symbol}*1\dfrac{MSymbol}{s}=2\dfrac{Mbit}{s}$ | $11\dfrac{chips}{symbol}*1\dfrac{MSymbol}{s}=11\dfrac{MChips}{s}$ |
| 5.5 Mbps | 8 chips / bit<br>4 bits / symbol<br>1.375 symbol/s | $4\dfrac{bit}{symbol}*1.375\dfrac{MSymbol}{s}=5.5\dfrac{Mbit}{s}$ | $8\dfrac{chips}{symbol}*1.375\dfrac{MSymbol}{s}=11\dfrac{MChips}{s}$ |
| 11 Mbps | 8 chips / bit<br>8 bits / symbol<br>1.375 symbol/s | $8\dfrac{bit}{symbol}*1.375\dfrac{MSymbol}{s}=11\dfrac{Mbit}{s}$ | $8\dfrac{chips}{symbol}*1.375\dfrac{MSymbol}{s}=11\dfrac{MChips}{s}$ |

*Figure 4.33: Overview of the four transmission procedures*

For this reason higher data transfer rates can only be achieved, with the same output power and antenna characteristics, if the channel quality is good enough. If the channel quality sinks below the required threshold the transfer can still take place at a slower speed. Most devices can make this type of adaptive change to the transfer speed.

*Table 4.8: Requirements placed on the SNR (signal strength-to-noise) ratio for various speed levels as defined in the 802.11b standard*

| Data rate | SNR requirement, relative to 1 Mbps |
|-----------|-------------------------------------|
| 1 Mbps    | 0 dB                                |
| 2 Mbps    | 2 dB                                |
| 5.5 Mbps  | 4 dB                                |
| 11 Mbps   | 8 dB                                |

The IEEE802.11b standard defines two packet frame coding options for higher speeds. One is to enter the higher speeds in the signal field of the normal PLCP-PPDU, as shown in Figure 4.29. (There are also some other small changes that are not described in any more detail here). However, even if you do this, the control frame is still transferred at 1 Mbps. One of the effects of this is that the protocol takes up proportionately more of the transfer.

To minimise this increase the standard also has a shorter format which halves the header transfer time by doing the following:

- halving the length of the PLCP preamble from 144 to 72 bits
- increasing the PLCP header's transfer speed to 2 Mbps without changing its length.

This also means that the shortened header is still compatible with IEEE802.11 systems which only support data rates of 1 and 2 Mbps.

### 4.7.3   Other standards

Alongside the extensions to standards 802.11a and b mentioned above there are a number of other activities that are designed to help the success of the standard itself. These are listed in Table 4.9:

*Table 4.9: Extensions to the IEEE802.11 standard*

| Name of Task Group | Description | Status |
|--------------------|-------------|--------|
| 802.11a | Definition of a PHY using an OFDM modulation process with 10 carrier frequencies at 5 GHz and data rates of up to 54 Mbps. | Released as IEEE Standard 802.11A-1999 |
| 802.11b | Definition of a PHY using a DSSS modulation process with 10 carrier frequencies at 2.4 GHz and data rates of up to 11 Mbps. | Released as IEEE Standard 802.11b-1999 |
| 802.11b-cor1 | Corrections to management information base (MIB) in 802.11 | In progress |
| 802.11c | Addition of specific MAC procedures to IEEE802.11, to enable 802.1-compatible bridging functionality. | Ratified as part of the ISO/IEC 10038 (IEEE 802.1D) standard. |

| Name of Task Group | Description | Status |
|---|---|---|
| 802.11d | Extension of PHY definitions (for example, channel selection, frequency hopping sequences, management information base (MIB) attributes) to apply to new countries (*regulatory domains*). | In progress |
| 802.11e | Improvements in service support (Quality of Service, Class of Service) for transfer efficiency in the Distributed Coordination Function (DCF) and Point Coordination Function (PCF), as well as for security mechanisms (released since May 2001 in 802.11i, see below). | In progress |
| 802.11f | IAPP (*Inter Access Point Protocol*) for *roaming* and *Load Balancing* | In progress |
| 802.11g | Extension to the 802.11b standard for 22 Mbps | In progress |
| 802.11h | Extension of the 802.11 MAC protocol and the 802.11a PHY protocol in the 5GHz range: selection of indoor and outdoor channels, improved channel monitoring and power management in agreement with CEPT and other European governing bodies | In progress |
| 802.11i | Extension of the MAC protocol to improve security and authentication mechanisms. Adoption of security mechanisms from 802.11e May 2001 | In progress |

*Table 4.9: Extensions to the IEEE802.11 standard, continued*

Some of the extensions to the standard which have not yet been ratified are briefly described below:

- In the case of devices that comply with 802.11 and 802.11b, and are designed for use in the 2.4 GHz band, you should be aware that there are no globally-accepted regulations for using these frequencies, even if the discussion here is not so heated as talks about 802.11a devices. So-called "world mode", 802.11d, allows the transmission of country-specific information from an access point to mobile PC adapter cards, which can then use this information to modify the characteristics of their radio transfer.

- 802.11e also attempts to extend the MAC layer so that data streams can be transported using increased functionality resulting from the implementation of PCF. Support for PCF for roaming and peer-to-peer operation is still fraught with difficulties.

- The operation of more complex networks is also discussed in the context of 802.11f. These debates are focussed on communications between access points. In particular, the aim is to unify the exchange of management and access data.

- Among the other standards, 802.11g is of particular importance. It has been the subject of a dedicated working group since May 2000. Its goal is to set out a specification for a further data rate increase to 22 Mbps for the 802.11 MAC in the 2.4 GHz carrier band. The activities are carried out within IEEE802.11g by the *Higher Rate IEEE802.11b Study Group* (HRbSG). However, the seemingly endless negotiations also appear to be burdened by numerous political agendas. In July 2001 the OFDM modulation procedure (see also section 2.8, Orthogonal frequency division multiplex procedure) developed by Intersil seemed to have taken the lead over its Texas Instruments competitor, Packet Binary Convolution Coding (PBCC). However when this book went to press (Summer 2002) ratification had not been completed because OFDM had not received the number of votes it required in the first round of voting. Observers suspect that manufacturers do not want to hinder the market launch of IEEE802.11a systems by providing additional competition.

- To close the security loopholes discussed in section 10.1, the IEEE's Wireless Group *Task Group i* (TGi), was set up to work out improved MAC layer security in the specifications of 802.11i. However it must be said that at present the discussions in this group are controversial, and do not always proceed towards the aim, and therefore no procedure has yet been prepared for ratification. Nevertheless this governing body continues to introduce and discuss a great number of options.

# 5 Bluetooth

## 5.1 The standard

### 5.1.1 Background

Activities involving Bluetooth began back in 1994 when Ericsson's Mobile Communications Division produced a feasibility study on what could replace the many cable connections between mobile telephones and the various peripheral devices. Following that, the five companies IBM, Toshiba, Intel, Ericsson and Nokia jointly founded the *Bluetooth Special Interest Group* (BSIG) at the start of 1998, with the task of creating a non-proprietary standard for *Personal Area Networks* (PAN). At this time the aim was seen as something approaching *wireless USB*. The first version, 2.0, was finalised in 1999. The first update, Version 2.0b, came out in December 1999. The current version of the standard, 2.1, has been available since February 2002. It is generally seen as the first solid basis for products that really suit the market, as previous versions contained a range of inaccuracies and errors that resulted in errors in compatibility, in the clean implementation of pico networks and in clear master-slave assignment [Klinkenberg-Haas (1)]. Even today, Version 2.1 has a few weak points with regard to interoperability [Bläsner (1)]. The current version of the Bluetooth standard, V2.1, is available from *http://www.bluetooth.com/developer/ specification/specification.asp*.

The original small group of BSIG founder members has grown with the addition of a promoter group which first involved the companies 3Com, Lucent, Microsoft and Motorola. The BSIG now has around 2,500 members.

The technology is named after Harald II. Blaatand (Bluetooth), who was King of Norway between 940 and 981, and christianised the country, as well as unifying Norway and Denmark. His name is probably based on the two words "blå" (dark-skinned) and "tan" (big man), "big man" meaning "king" or "leader" in this context.

The Bluetooth standard is based on three application scenarios which position it at Personal Area Networks level:

- replacing cables
- data and voice access points
- personal ad hoc networks.

To provide these applications you do not necessarily need PCs. Often it is the case that relatively cheap and simple devices with a wireless connection can be used. For this reason the Bluetooth standard needs to allow the simplest and therefore cheapest solutions possible.

### 5.1.2  Structure

Figure 5.2. shows the main elements of version 2.1 of the Bluetooth standard, the current version.

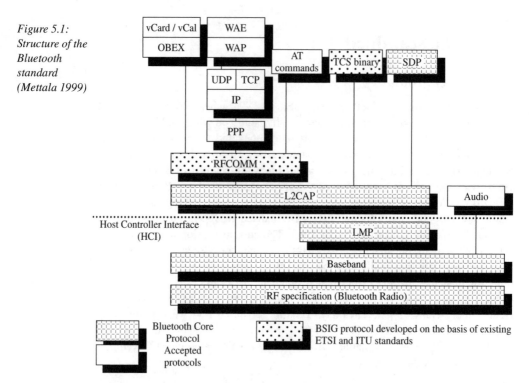

*Figure 5.1:*
*Structure of the*
*Bluetooth*
*standard*
*(Mettala 1999)*

Here it is worth noting the following:

- the description of the Bluetooth communications protocols includes all levels of the protocol layers. This not only includes the control protocols for creating ad hoc networks, but also the elements for transferring isochronous data streams, all of which go far beyond the purely transport-specific services of an IEEE802.11 network. Bluetooth provides a complete radio system. Although this is an advantage it must be viewed against the background of interoperability between different device types (see section 5.7).

- the complete protocol stack does not only contain Bluetooth-specific protocols: it also uses existing and extended protocols, especially in the higher application-oriented layers. For example it includes the Internet protocols defined in the TCP/IP stack.

- when you consider the large number of protocol elements shown in Figure 5.1 it becomes immediately obvious that a Bluetooth station that completely implemented all of them would be so complex that it would be an obstacle to the aim of simple and cheap implementation. For this reason, instead of a complete implementation, the plan is usually to implement only the necessary elements in devices. Each station only needs to contain the core protocols.

- the Bluetooth specification also includes the Host Controller Interface (HCI) which provides a command interface to the baseband controller and link manager, and to the hardware status and command registers. The positioning of the HCI can be modified. Figure 5.1 shows a typical example.

- when the standard was defined no attempt was made to ensure conformity with the layers in the extended reference models (OSI reference model or TCP/IP reference model).

- Bluetooth was not defined in an existing standards committee. Efforts have been made, especially by the IEEE, to fit Bluetooth into other standards families, such as the standards defined in IEEE802.x, but as the points above make clear, Bluetooth will need considerable modification before this is possible. The IEEE802.15 attempts to describe the PHY and MAC-portion of the Bluetooth specification.

### 5.1.3    Standard Bluetooth elements

Below you will find a brief introduction to the most important elements of the Bluetooth standard.

- RF and baseband: these two specifications correspond approximately to the PHY and MAC layers in the IEEE protocol stack. The sections below describe how these two layers work.

- *Link Manager Protocol* (LMP): The LMP is responsible for network management tasks. In particular these include creating connections between stations, authentication and encryption, controlling power-saving modes and monitoring the status of devices in a pico network.

- *Logical Link Control and Adaptation Protocol* (L2CAP): L2CAP connects the protocols of the higher layers with the tasks of the baseband. It can be said that L2CAP is arranged in parallel with LMP because it is responsible for transferring useful data. Although L2CAP provides both connection-oriented and connectionless services it only uses the connectionless asymmetrical ACL connections defined in the baseband protocol.

- *Service Discovery Protocol* (SDP): the SDP recognises different services and the characteristics of each service. It is an essential element of the Bluetooth protocol. With it you can find out which services are available and use that information to create a connection. SDP is the basis for Bluetooth stations to create relatively simple ad hoc networks.

- *Wireless Cable* (RFCOMM): the RFCOMM standard is part of ETSI specification TS. 07.10. It has been modified to suit the Bluetooth standard, and emulates a serial (RS-232) port. It is essential for many applications and is a clear indication of the intention and range of services of the Bluetooth standard. It primarily describes a point-to-point connection between two devices over the "air serial port". For this reason, for example, LAN connections via TCP/IP are also implemented using the serial Point-to Point protocol (PPP) via RFCOMM.

- *Telephone coupling:* Two elements describe how telephone services are to be coupled. *Telephony Control Service – Binary* (TCS Binary) specifies the signalling for establishing and ending calls. It is based on ITU-T recommendation Q.932. The Bluetooth SIG uses a set of AT commands from the appropriate ITU and ETSI specifications to provide a means of addressing modems and mobile telephones. Fax services are also available.

  Note that only call establishment is carried out using the TCS-BIN protocol and the L2CAP layer. The audio protocol is used to transfer useful data. This protocol directly accesses an SCO link in the baseband.

In addition to this, various application-related protocols have found their way into the Bluetooth standard:

- One is the *Object-Exchange* (OBEX) protocol, which was adopted from the IrDA standard for infrared connections. It models the representation of objects and the structuring of dialogs between those objects.

- OBEX is used to handle the *vCard* protocol with which virtual business cards can be exchanged.

- The *Wireless Application Protocol* (WAP), was developed for transmitting WML resources on mobile telephones. It is also available in the Bluetooth standard, and is implemented via the TCP/IP protocol stack.

## 5.1.4   Profiles

The Bluetooth standard also contains details of each protocol stack. In particular you will find definitions called "profiles" that specify the way in which application software can access the various layers of the protocol stack. In most cases this involves *Application Program Interfaces* (APIs). These cover the majority of possible application models and correspond to the service access points in the ISO/OSI reference model (see section 2.2.1). This has the advantage that programs from different manufacturers on different devices can communicate with each other without difficulty as long as they comply with these definitions, which are as detailed as possible. However, so much is involved that these definitions are highly complex. As a result, these three points need to be made:

- Creating complete and consistent definitions is very time-consuming and has been the main aim of revisions to the standard up to the current version, V2.2.

- The standard is becoming very extensive. For example, version V2.1 includes 13 application-related profiles for activities such as communicating via a *serial port*, a headset, or a fax-enabled device, for connecting to a LAN, and for data transfer via *File Transfer Protocol* (FTP).

- Nevertheless, even error-free and efficient use of the interfaces is not possible without extensive know-how, which initially makes it more difficult to create new Bluetooth developments.

## 5.2    Architectures

A Bluetooth network essentially uses a master/slave architecture in which the master controls the traffic flow. In this way relatively simple isochronous traffic streams can be handled, as found, for example, in the audio module. All Bluetooth devices have identical hardware properties so that the master is only selected when the network is established. Section 5.4.2 lists the criteria that decide which is chosen as the master.

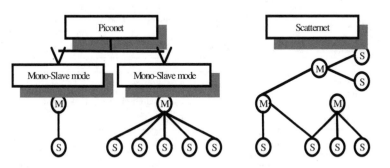

*Figure 5.2: Architectures of Bluetooth networks*

As Figure 5.2 shows, the following types of Bluetooth network exist:

- If only one master is present in a network then the network is a *pico network*.
  - If a master only uses a point-to-point connection to communicate with the slave then the network is operating in *mono-slave mode*.
  - However the master can also operate in *multi-slave mode*, establishing point-to-multipoint connections to seven active slaves, while additional slaves can participate passively in a "parked" state. Section 5.4. describes the different types of state used.
- A *scatternet* consists of several combined or *scattered* pico networks. As each pico network is managed by its own master this means that there are several masters in scatternet mode. In that mode a station can be logged on as a slave in several pico networks and then also be active in these networks. At the same time, the master of a pico network can also be a slave in another pico network.

## 5.3    Channel access

### 5.3.1    Prioritisation

The rules defined in the Bluetooth standard differ from the rules in IEEE802.11 in crucial ways. For example, in Bluetooth, different traffic types are explicitly meant to be supported with a guaranteed service. Unlike in PCF, Bluetooth does this by precisely defining the times at which the master and slaves are permitted to transmit. The master manages the selection of slaves to permit them to access the channel at a particular time, centrally. This provides a simple way of managing transmission capacity.

To fulfil the different requirements of the various traffic types Bluetooth provides various mechanisms with the help of connections, packets and channels. These are listed in Figure 5.3, and described below.

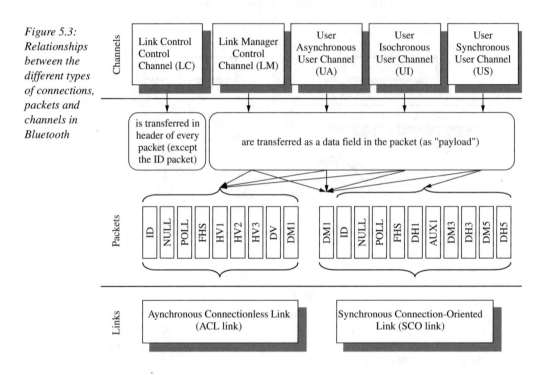

*Figure 5.3: Relationships between the different types of connections, packets and channels in Bluetooth*

### Connections

The previous versions of the Bluetooth standard defined two basic communication types:

- Synchronous connection-oriented communication via a *Synchronous Connection-Oriented Link* (SCO) creates symmetrical point-to-point communication between the master and a single slave. In functional terms SCO corresponds to a circuit-switched transmission as the master regularly reserves time slots. This means that the master sends data to the slave in fixed time slots and the slave is then authorised to sends its data in each subsequent time slot. The interval between two transmissions from the master to the slave is called the $T_{SCO}$.

A master can establish up to three SCO links to one or more slaves. A slave, on the other hand, can establish up to three SCO links with one master or support up to two SCO links from different masters.

SCO links are designed to ensure efficient voice transmission. As section 2.3 showed, a certain amount of data loss is non-critical when voice data is being transmitted. What is much more noticeable, and disruptive, is any delay in transmission. For this reason no data integrity checks are made on SCO links. If data is lost during transmission it is also not retransmitted because this would again cause a delay for the voice data that follows.

- In contrast, asynchronous connectionless communication via an *Asynchronous Connectionless Link* (ACL) creates a link between the master and one or more slaves. This link can only be created if the channel is not reserved for an SCO. An ACL link corresponds to a packet-switched transmission. Only one ACL link can be established between a master and a slave at any particular time.

When ACL links are in use the master can also send packets to all the slaves in its pico network by not specifying a target address. These packets are then interpreted as a broadcast. Slave broadcasts are not supported.

Unlike SCO links, ACL links are designed to transmit data efficiently. When data is being transmitted, data integrity is usually far more important than delays. As a result, when ACL links are in use, incorrect or missing information blocks are requested again, and retransmitted.

### Bluetooth packet format

In each of these connection types various different packet types can be sent. Figure 5.4 shows an overview of them with their most important characteristics. The various types are coded in the *Type* field in the packet header shown in Figure 5.5, with the help of 4 bits.

| | Type | Payload Header [Bytes] | User Payload [Bytes] | FEC | CRC | Symmetric Max Rate [kbps] | Asymmetric Max Rate [kbps] | |
|---|---|---|---|---|---|---|---|---|
| | | | | | | | Forward | Reverse |
| Link Control Packets | ID | - | - | - | - | - | - | - |
| | NULL | - | - | - | - | - | - | - |
| | POLL | - | - | - | - | - | - | - |
| | FHS | - | - | - | - | - | - | - |
| ACL Packets | DM1 | 1 | 0 .. 17 | 2/3 | yes | 108.8 | 108.8 | 108.8 |
| | DH1 | 1 | 0 .. 27 | no | yes | 172.8 | 172.8 | 172.8 |
| | DM3 | 2 | 0 .. 121 | 2/3 | yes | 258.1 | 387.2 | 54.4 |
| | DH3 | 2 | 0 .. 183 | no | yes | 390.4 | 585.6 | 86.4 |
| | DM5 | 2 | 0 .. 224 | 2/3 | yes | 286.7 | 477.8 | 36.3 |
| | DH5 | 2 | 0 .. 183 | no | yes | 433.9 | 723.2 | 57.6 |
| | AUX | 1 | 0 .. 29 | no | no | 185.6 | 185.6 | 185.6 |
| SCO Packets | HV1 | - | 10 | 1/3 | no | 64.0 | - | - |
| | HV2 | - | 20 | 2/3 | no | 64.0 | - | - |
| | HV3 | - | 30 | no | no | 64.0 | - | - |
| | DV* | 1 D | 10 V +0 .. 9 D | no (V) '2/3 (D) | no (V) yes (D) | 64.0 (V) + 57.6 (D) | - | - |

* D stands for "data field", V for "voice"

*Figure 5.4: Different Bluetooth standard packet types and ways in which they can be combined*

- *Common packet types:* the *common packet types*, ID, NULL, POLL and FHS packets, can be used for both asymmetric and symmetrical links. Their main purpose is to manage participant stations.

- *DM (Data – Medium Rate)* packets carry data via a transfer speed change request. The number 1, 3 or 5 represents the number of time slots that the packet adopts (see also section 5.5). This also applies to the packet types that follow. A DM1 packet can be sent in both connection types and is a kind of basic connection.

  In particular, during a synchronous connection it is also possible to transmit control information.

- *DH (Data – High Rate)* packets carry data via a high transfer speed change request. The only difference between them and DM packets is that they do not use FEC redundancy.

- *HV (High Quality Voice)* packets are designed to transfer speech and other synchronous and transparent services.

- *DV (Data Voice Combined)* packets consist of a data block and a speech block. The two parts are processed entirely independently of each other. This also applies to their error handling through encoding and redundant data transmission.

## 5.3.2    Frame formats

The basic bit sequence of a Bluetooth packet is shown in Figure 5.5 with some essential notes. A particularly interesting feature is the access codes, which are used to manage the stations in the network.

*Figure 5.5: Bluetooth packet format*

| Access Code | | | Header | | | | | | Payload |
|---|---|---|---|---|---|---|---|---|---|
| | | | Link Control (LC) information | | | | | | |
| 72 b | | | 54 b | | | | | | 0 to 2745 b |
| Pre-amble | Sync Word | Trailer | AM_ADDR | Type | Flow | ARQN | SEQN | HEC | |
| 4 b | 64 b | 4 b | 3 b | 4 b | 1 b | 1 b | 1 b | 8 b | |

Pre-amble / Sync Word / Trailer: *There are three access codes*
- *CAC Channel Access Code*
- *DAC Device Access Code*
- *IAC Inquiry Access Code*
  - *GIAC General Inquiry Access Code*
  - *DIAC Dedicated Inquiry Access Code*

AM_ADDR: *Active Member Address*

Type: *Packet type*

Flow: *for ACL flow control: the receiver sets it to STOP (= 0) to inform the transmitter that transfer is to be split.*

ARQN: *Acknowledgement Indication. ARQN is transferred in the header of the reply packet and signals the error-free reception of a packet with reference to the CRC code.*

SEQN: *Sequential Numbering. SEQN is inverted in every packet transferred with CRC.*

HEC: *Header Error Check*

Figure 5.5 does not show the useful data field (*Payload* field), which always contains 240 bits, for the symmetrical voice packets, and therefore has a very simple structure. However, the data packets, as sent over asymmetric ACL links, must have a structured format, due to their high levels of flexibility. Figure 5.6. shows this structured format.

Note the way that payload header can differ in length according to the number of time slots in the packet. The number of time slots used depends entirely on the length of the packet that requires coding. Also note the assignment to the logical channels of the upper protocol levels, L2CAP and LMP (see also section 5.1.2), which are identical for both single-slot and multi-slot payload headers.

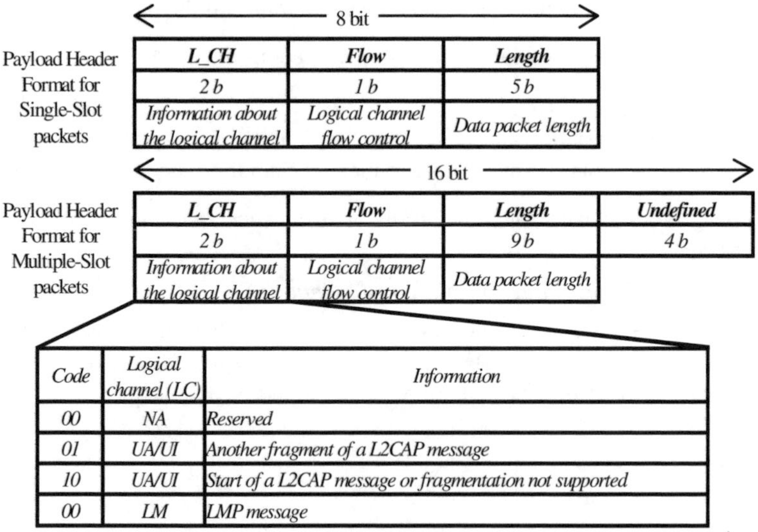

*Figure 5.6: Format of the Bluetooth data packet*

| Code | Logical channel (LC) | Information |
|------|----------------------|-------------|
| 00 | NA | Reserved |
| 01 | UA/UI | Another fragment of a L2CAP message |
| 10 | UA/UI | Start of a L2CAP message or fragmentation not supported |
| 00 | LM | LMP message |

### 5.3.3    Audio transmission

The symmetrical and synchronous transmission of audio data is an essential element of the Bluetooth standard. As telephone applications take centre stage here, several CODECs (*Coder/Decoder*) have also been specified.

- A CODEC is based on a logarithmic *Pulse Code Modulation* (PCM) procedure with two characteristics (A-law and μ-law). This is also used for a data rate of 64 kbps in ISDN telephony.

- A *Continuous Variable Delta Modulation* (CVSD) CODEC transfers the data with delta modulation.

## 5.4    Controlling states

### 5.4.1    Introduction

In a Bluetooth network, managing stations is especially important because each station must be in a defined state at any specific time so that transfers initiated by the master can be triggered in the pico network. Here, the communications mechanisms for establishing a network and the rules for processing communication in an established network are completely different. When a pico network is established the master is also specified at the same time, and that master centrally manages access to the channel. On the other hand, when a network is closed, a decentral access mechanism must still be present.

The Link Controller in a Bluetooth station can adopt the states shown in Figure 5.7.

Standby state is the initial state of each Bluetooth station on power-up, and also permits low-power-operation. The station leaves Standby state

- to process a received page call (*Page Scan*) or received inquiry call (Inquiry Scan) or
- to send a page call (*Page*) or inquiry call (*Inquiry*).

Once a page call has been successfully processed the station changes to the *Connection* state. Once the station has replied to a page call it is operated as a slave. If the station itself has sent the page call it acts as the master.

*Figure 5.7:*
*Bluetooth link*
*controller*
*states*

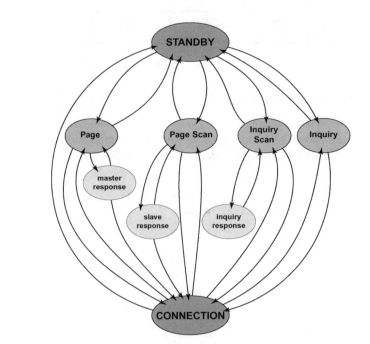

## 5.4.2    Establishing a connection

### Overview

When a connection is being established *Inquiry* and *Paging* take on two separate tasks:

- Using an *Inquiry* stations can find out which other stations are within transmission range, and what their addresses and clock states are.
- Stations can use *Paging* to establish a connection between them.

Each station that is switched on sends inquiries to the environment around it at regular intervals, and monitors the channel in the times between two of its own inquiries for inquiries from other stations.

### Inquiry

A Bluetooth station that wants to get information about other stations in its environment sends an Inquiry message. This is an ID packet that

- either contains a *General Inquiry Access Code* (GIAC), if all stations present are to answer,
- or it contains a *Dedicated Inquiry Access Code* (DIAC), with the number of a particular device class, if only certain types of devices are to answer.

If a station receives an Inquiry, it changes to the *Inquiry-Scan* state by responding with an FHS packet that contains its device address (*Device ID*). The station can only reply to an Inquiry if it is a slave. The station that sends the Inquiries becomes the master. When it receives an answer to its Inquiry it adds it to a list. However, it does not send an acknowledgement to the slave. This is important because, when several slaves are replying to an Inquiry, they may do so simultaneously or almost simultaneously. After all, the pico network is not yet established, the stations are therefore not yet synchronised, and the master has still not yet begun managing channel access. Nevertheless, Bluetooth does also implement a random backoff algorithm, which is very similar to the algorithms defined in IEEE802.x, to minimise the probability of simultaneous Inquiry answers. If a collision occurs the Inquiry answers from both slaves are lost. Accordingly, they cannot answer until they receive the next Inquiry.

Another problem that arises in the organisation of the Inquiry algorithm is that stations

- are still unaware of each other. In particular, they do not know the device identification numbers which are the basis for calculating the frequency hops. As a result, a fixed hopping sequence called *Inquiry* is used for the Inquiry, and another fixed hopping sequence called *Inquiry-Response* is used for the answer.
- are not yet synchronised. As a result, the use of pre-defined frequency hopping sequences does not guarantee that stations will also understand each other.

Bluetooth stations continue with the Inquiry process shown in Figure 5.8 until they have detected a pre-defined number of stations or a pre-defined time interval has passed.

Figure 5.8:
Inquiry
algorithm used
by Bluetooth
stations

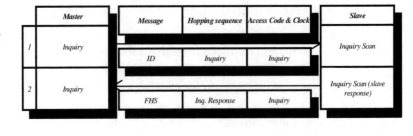

### Paging

The master uses the *Page* state to establish a connection with a slave. Here, the master attempts to access the slave by repeatedly sending its **Device Access Code** (DAC). However it can only reach the slave if the slave is in the *Page Scan* state, which it adopts regularly. The difficulty is that the master knows neither the time nor the frequency hopping sequence at which the slave "wakes up". As a result, the master uses a cunning algorithm to ensure it is received by the slave after as few attempts as possible.

Once a slave has successfully received a Page inquiry it returns a Page Response to which the master then replies in turn with an FHS message. The information this message contains includes, importantly, the master's *Device ID*, from which the frequency hopping sequence can later be derived for data transfer, and the clock state. Once the slave has sent an acknowledgement, in the form of a new Page Response, the transfer can begin. Once communication has been established between the master and slave, in the connection establishment phase, using the clock state and the slave's Access Code, the transfer can take place, as defined by the master. Figure 5.9 provides an overview of this process.

*Figure 5.9:
Page algorithm
used by
Bluetooth
stations*

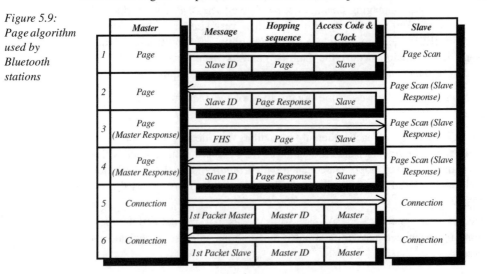

### 5.4.3    Operating a connection

Once the connection has been successfully established, using Paging, both the master and the slave station adopt the *Connection* state. Slaves can adopt any of the state modes shown in Figure 5.10. These are used for two reasons: to minimise power consumption and to allow more complex networks, based on scatternets, to be established.

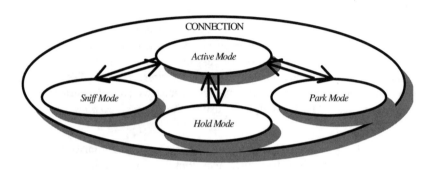

*Figure 5.10: States adopted by a Bluetooth controller when operating a connection*

### Active Mode

In Active mode both the master and slave actively use the transmission channel. Active slaves expect packets from the master in even-numbered master slots. If the master does not send a packet, or sends a packet to another slave, the slave can enter sleep mode when the master sends its next transmission. In particular, the slave can use the information in the control section to also find out if the master is sending a packet that contains three or five slots. The slave then waits until it later receives the next transmission containing four or six slots from the master.

### Sniff Mode

In sniff mode the frequency at which the slave is ready to receive packets from a master reduces. As the master can therefore no longer address the slave in every master slot, the master and slave must negotiate the time interval at which the slave is periodically ready for reception. To do so, either the master or slave sends a sniff command as defined in the LM protocol.

Once the time period $T_{sniff}$ has expired, the slave waits to receive packets from the master for a certain number of slots, defined in $N_{sniff\,attempt}$. If the slave receives no packets addressed to it, in this time period, it goes to sleep for the period defined in $T_{sniff}$. However, if it detects a packet with the right address, it is then also ready to receive more packets immediately after. It is therefore not possible for a slave to start entering sleep mode during transmission in sniff mode.

Sniff mode can also be used for another purpose: to be a member of two pico networks, in a time multiplex, when there is a scatternet.

**Hold Mode**

For longer periods of time during which the slave is not to participate in ACL-based communication, the slave enters "hold mode". If it does so it stores the parameters for the connection, especially the addresses, stored in the *Active Member Address* (AM_ADDR) and synchronisation information. In this hold mode, however, it still needs to be able to receive any SCO packets.

**Park Mode**

If the slave is in "park mode", it does not participate in communications in the pico network at all. To arrange this it is provided with two addresses with which the master can address it to bring it out of park mode again. All slaves that are in park mode are assigned the *Parked Member Address* (PM_ADDR), which contains eight zeros and is accordingly used as a broadcast address. Furthermore, the slave can leave park mode by using the *Access Request Address* (AR_ADDR).

In park mode the slave wakes up at regular intervals to synchronise itself and monitor the channel for broadcast messages. For communication with parked slaves the master uses beacons.

In park mode the slave uses less power, and the use of park mode also allows the pico network to include more than seven slaves.

### 5.4.4    Addressing

In the Bluetooth standards addresses occur at several levels:

- The *Bluetooth Device Address* (BD_ADDR), length 48 bits, is permanently assigned to the devices in accordance with the format defined in IEEE802.
- As long as active slaves are participating in a pico network they are assigned a 3-bit *Active Member Address* (AM_ADDR). Again, the broadcast address consists of zeros (nulls). In this way broadcast messages can be sent to all the participants in a pico network. An important point is that these broadcast messages are also received by the parked stations.
- The *Parked Member Address* (PM_ADDR) and the *Access Request Address* (AR_ADDR) are used to handle park mode, as described in section 5.4.3.
- In addition, the Bluetooth system also uses access codes. These form the first section of the frame shown in Figure 5.5, and are, to some extent, derived from the addresses.

## 5.5    Bit transfer

Like others, the Bluetooth standard is also designed for use in the 2.4 GHz ISM band and must use a spread spectrum technique as defined in the regulations. To do so, the Frequency Hopping Spread Spectrum process (FHSS) is used, as it is simpler and cheaper to implement than DSSS. In version 2.x, however, it limits the data rate to 1 Mbps.

In addition, regional restrictions also need to be taken into consideration (as is already the case in the IEEE802.11 standard: see section 4.4). Costs and effort are minimised by using only two operating modes: one mode with 79 hopping frequencies, for North America and Europe (except France and Spain), and a second mode with 23 hopping frequencies, for Japan, France and Spain.

The frequency sequence is specified by the master in the established pico network and follows a pseudo random number sequence which is calculated, using relatively complex rules, from the master's device address, and is therefore unique to each pico network. This supports the operation of the maximum possible number of independent pico networks in a small geographical area.

The nominal hop rate is 1600 hops/s. To set this up the master assigns all slaves in the pico network time slots with a length of 625 µs. Participants can only start transmitting at the start of a time slot. The time slots are numbered from 0 to 227-1, depending on the master's clock, so there is a counting cycle time of 23 h.

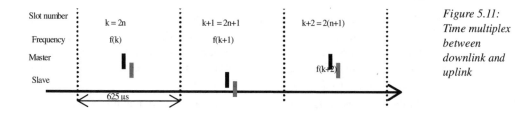

Figure 5.11:
Time multiplex
between
downlink and
uplink

To support bi-directional traffic a *Time Division Duplex* (TDD) process has been implemented in which the master is only permitted to begin its transmission at the start of an even-numbered time slot, and slaves are only permitted to begin their transmission at the start of odd-numbered time slots. This is shown in Figure 5.12.

Two packets that are next to each other are transmitted at different frequencies. In normal mode a frequency hop is made at the start of each new time slot. When a packet containing three or five time slots is transmitted, the same frequency is used until the entire packet has been transmitted. The packet that follows will be transmitted using the frequency that corresponds to the clock state. Figure 5.12 shows this change in frequency.

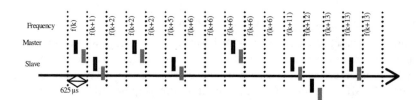

Figure 5.12:
Transmitting a
packet that
requires
several time
slots

This cycle time can also be used to derive the requirements for isochronous voice transmission. A packet of type HV1 contains 10 bytes = 80 bits of useful data. Assuming a required data rate of 64 kbps for ISDN-quality voice transmission, an HV1 packet can be used to transmit 2.25 ms of speech.

If we use the time slot period of 625 s, this means that an HV1 packet needs to be transmitted in every second time slot. This is known as $T_{SCO} = 2$. If there were symmetrical transmission, in which data were transmitted in the feedback channel at a data rate of 64 kbps, this would mean that a channel was completely busy. This example is a very impressive way of showing how little performance a channel has if it has a gross data rate of 1 Mbps. The transmission efficiency in this case is around 12.8 % (=128 kbps/1 Mbps).

## 5.6   Security

Bluetooth uses similar processes to the ones defined in IEEE802.11 to provide the security measures necessary in wireless systems. Three levels of security are implemented:

- The Bluetooth Device Address (BD_ADDR), whose length is 48 bits, corresponds to the MAC address assigned by IEEE802. This address, which is unique for every station, is publicly recognised. If the station is being operated as a master this Bluetooth Device Address is used to calculate the frequency hopping sequence for use in the pico network.

- At the next level a 128-bit *authentication key* is used to authenticate stations.

- A key is also used to protect the information that is transferred. However, once again, the length of the code can be varied from 8 to 128 bits to allow the production of the most cost-effective systems possible. Here the *encryption key* can be derived from the authentication key.

The mechanisms for logon and encryption are similar to those used in IEEE802.12.

## 5.7   System implementation

Compared with the other technologies described in this book the demand for Bluetooth system implementation is the most significant because, in it, the higher protocol layers have been involved more than in those technologies. This allows the partitioning of Bluetooth implementation to be divided up along the communications path. The main components can easily be identified in the protocol stack shown in Figure 5.1:

- HF part with antenna, transceiver and modem

- Low-level baseband controller with baseband processing, and control and management of the connection up to the Host Controller Interface (HCI).
  A 32-bit microprocessor is usually used to carry out baseband processing. The ARM7TDMI processor is frequently used for this purpose [Bläsner 2001 (1)].

- Higher-layer protocol stacks

- Application software.

This set of elements can be combined to form a number of practical system configurations. They are based on two main application scenarios, shown in Figure 5.13.

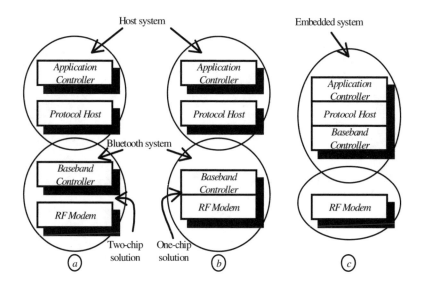

*Figure 5.13:*
*System*
*configurations*
*of a Bluetooth*
*module*

- In the case of host-based applications the Bluetooth module only provides the communications functionality (layers 1 and 2 in the reference model), and the application-related functionality can be implemented in a flexible way by software on the host. Here it may be appropriate to use a single-chip solution (configuration *b*), if no special requirements are placed on the RF module, and monolithic integration using cheap CMOS-based processes is possible [Sikora 2001 (5)]. If this is the case, the two-chip solution (configuration *a*) will take precedence for the foreseeable future.

- If Bluetooth is being implemented in an embedded system that is only designed to handle one application, or a small number of applications, it would be appropriate to provide all the functionality involved on one chip. However, single-chip solutions are not usually able to hold the necessary RF components as well. This is the reason for configuration *c*.

In this configuration it is useful to use ready-made design blocks along the lines of IP (which in this case, unusually, does not mean "Internet Protocol", but *Intellectual Property*).

# 6 DECT

## 6.1 The standard

The DECT standard ETS 300 175 for *Digital European Cordless Telecommunications* was defined by the European Telecommunications Standardization Institute (ETSI) in 1992. Since then there has been a dramatic increase in the use of wireless telephones by both domestic and commercial users in more than 100 countries, especially almost every major industrialised nation. The development of a European protocol standard for the local connection of portable voice telephones to a fixed base station that is digital, secure against eavesdropping, stable and easy to use, was and still is a complete success. As an alternative, the *Personal Handy-Phone System* (PHS) has established itself in Japan and some other East-Asian states. The two standards have many features in common.

Due to its extraordinarily successful market launch, estimated at 38 million units world-wide in 2001 (*www.dectweb.com/introduction_statistics/.htm*), and a total installed base of around 100 million units, DECT has the immense advantage of the large number of radio modules that have already been installed, and are therefore cheap to access. On-going upwards integration [Hascher] and new technology in the HF module [Bläsner (2)] promise continued reductions in costs.

To ensure that DECT data products were compatible with each other the companies Ascom, Canon, Dosch & Amand, Ericsson, Hagenuk and National Semiconductor set up the DECT Multimedia Consortium (DECT-MMC), *www.dect-mmc.com*, at the DECT World Congress 1999 in Barcelona, Spain. Since then more companies have joined the initiative.

You can find more detailed information and notes on the DECT website (*http://www.dectweb.com*) and the DECT forum (*www.DECT.ch*). You will find a general overview of the DECT technologies in [Sikora (2)].

## 6.2 Architectures

DECT is designed to handle a point-to-point connection between a base station, the *Fixed Part* (FP) and a mobile device, the *Portable Part* (PP). Here, a base station can process the traffic between several mobile devices and the fixed network, or also the mobile devices can do so themselves. In addition, several base stations can be interconnected to allow roaming and handover.

## 6.3   Channel access

An MC/TDMA/TDD algorithm is used to distribute the frequencies for the various channels within this frequency band. This means:

- *Multi Carrier* (MC): that several carrier frequencies are available in the form of a frequency multiplex (see also section 2.6.3)

- *Time Division Multiple Access* (TDMA): that within a carrier frequency several time slots can be used for the various channels, one after the other, in a similar way to a normal time multiplex (see also section 2.6.2)

- *Time Division Duplex* (TDD): that the downlink transfer between the base station and mobile device and the uplink transfer between the mobile device and base station can both be multiplexed, and transmitted on one carrier frequency, using different time slots (see also section 2.6.2).

*Figure 6.1:*
*Frequency/time*
*spectrum in*
*DECT*

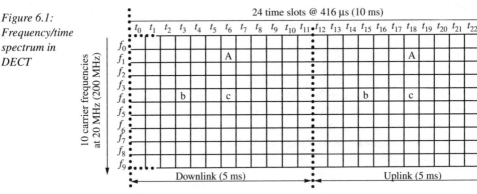

Figure 6.1 shows the DECT frequency/time spectrum. 24 time slots, with a total duration of 10 ms, are available on 10 different carrier frequencies. Twelve of these time slots are used for downlink transfer from the base station to the mobile device, and the other twelve for the uplink from the mobile device to the base station. In Figure 6.1, three transfers are active:

- Connection **a** uses every seventh time slot ($t_6$ for the downlink and $t_{18}$ for the uplink) on the second carrier frequency ($f_1$)

- Connection **b** uses every fourth time slot ($t_3$ for the downlink and $t_{14}$ for the uplink) on the fifth carrier frequency ($f_4$) and

- Connection **c** uses every seventh time slot ($t_6$ for the downlink and $t_{18}$ for the uplink) on the fifth carrier frequency ($f_1$).

In total there are therefore 120 channels that can operate in parallel, without any interference, within a cell. The bandwidth of a channel is 32 kbps for ADPCM-coded voice transmission. The total bandwidth available per carrier frequency and direction of transfer, including the control time slots that are not shown in Figure 6.1, is 522 kbps .

## 6.4    Bit transfer

DECT works in a reserved frequency area which, in Europe, lies between 2.88 and 2.9 GHz. On other continents other frequency ranges of 2.5 GHz to 3.6 GHz are used, and in a few cases the publicly-released 2.4 GHz ISM band is also used (see also section 2.11). The DECT technologies that are used in these higher frequency areas are also called *Upbanded DECT*.

The base station selects each transmission channel in a background process by regularly evaluating the activity on all channels in its reception zone. The free channels are then stored in the RSSI (*Received Signal Strength Indication*) list according to the activity on them. Then, if there is a request to create a new connection, the base station can select the channel that is best suited (least disrupted) for that connection.

The DECT mobile device (Portable Part) in turn also continuously monitors the activities in the relevant frequency areas and determines if the signals come from a base station to which the mobile device must log on. The mobile device then logs on to the base station that has the most powerful signal output power, to which it has the right to log on.

As these checks are also carried out dynamically during transfer (*Dynamic Channel Allocation and Selection*), this provides a flexible way to respond to disruption or change of location. For example, it is also possible to arrange the "handover" of a mobile device (Portable Part) that is on the move from one base station to another base station, while fully maintaining the connection, if the two base stations both permit logon, and are suitably linked (*Intercell Handover*). In addition, a mobile device can also change channel within a cell, while maintaining the connection, if the channel used up to that point is disturbed by a new event (*Intracell Handover*). This can become necessary, for example, if a mobile device that is transmitting in a cell comes physically close to another cell that is using the same channel for communication.

## 6.5    Application profiles

### 6.5.1    Voice telephony

The standard described above has been used to develop numerous other application-related standards which have extended the application area of DECT, especially in the area of voice telephony:

- A non-manufacturer-specific access profile (*Generic Access Profile* (GAP), ETS 300 444) ensures interoperability between base stations and mobile devices made by different manufacturers for voice applications.

- With the *DECT/GSM Interworking Profile* (GIP, ETS 300 370), GSM mobility functions can be used to couple different DECT domains, even if they are geographically separate.

- The *DECT Radio Local Loop Access Profile* (RAP): ETS 300 765) describes the implementation of public radio networks for connecting end customers with the help of DECT to increase competition, even for direct customer connections, while also keeping costs down. As numerous field tests in Germany, and initial installations in various other countries, including Italy and Poland, have shown, this is technically achievable. However, its economic effectiveness is seriously affected by the regulatory situation in each country, and the companies involved in the market.

## 6.5.2   Data transmission

The profiles described in the previous section are mainly designed to extend the functionality of voice telephony networks. In 2001 ETSI also finalised two more standards containing profiles for multimedia and network applications:

- The **DECT Packet Radio Service** (DPRS), ETS 301 649, adds the most important services for packet-oriented data transmission to the DECT standard. These especially include:

  - the negotiation and guarantee of particular services (*Service Negotiation –* SN) and

  - the dynamic distribution of resources (**Dynamic Resource Management –** DRM), to handle bursts of traffic efficiently. DRM also permits the bundling of several channels to achieve the higher bandwidths needed for multimedia applications. To bundle channels, the first step is to group the time slots in a carrier frequency. As data traffic is usually asymmetric it seems sensible to use both duplex transfer time slots for one direction of transfer so that bandwidths of up to 1 Mbps are available per carrier frequency. If the ten carrier frequencies are used simultaneously, this means theoretically that all 120 transfer channels can be used in parallel. By this means DRM aims to provide a bandwidth of up to 20 Mbps within a cell.

  The name DPRS is a deliberate reference to the *General Packet Radio Service* (GPRS) extension of GSM mobile telephony networks, as the procedure for bundling the time slots is identical.

- The *DECT Multimedia Access Profile* (DMAP), ETS 301 650, which is based on the *Multimedia Access Profile* (MMAP®) produced by Dosch & Amand, offers data services with added value services. DMAP is based on the GAP and DPRS standards already mentioned, but also includes extra services such as *Direct Link Access* (DLA) for ad hoc network connections.

# 7 HomeRF

## 7.1 The standard

The HomeRF standard was developed as an open company standard by the *Home Radio Frequency Working Group* (HomeRF WG) which was founded in March 1998 by Compaq, Hewlett-Packard, IBM, Intel and Microsoft. By Summer 2001 more than 50 companies had become members. The inner "promoters" group within HomeRF now includes the companies Siemens, Compaq, Motorola, National Semiconductor and Proxim. You will find the group's homepage at *http://www.homerf.org*. The standard is not publicly available.

There are currently two versions of the HomeRF standard that are of any practical significance. They are shown in Table 7.1.

| Version | Bandwidth | Status |
|---------|-----------|--------|
| V2.2 | 2.6 Mbps | 2000 |
| V2.0 | 10 Mbps | May 2001 |
| V2.1 | 20 Mbps | Currently being prepared for 2003 |

*Table 7.1: Versions of the HomeRF company standard*

As its name suggests, HomeRF is primarily intended for networking private households. However, the release of HomeRF version 2.0 also addresses smaller company networks: it not only supports fast speeds but also movement of stations between cells (*roaming*). HomeRF provides special support for the requirements made by different traffic streams with comparable data rates.

At the logical level HomeRF combines DECT's voice traffic technology with channel access for data traffic that resembles Ethernet. It also includes a third traffic type for multimedia data.

While the rules are being finalised you will sometimes also come across the name DECTplus [Tourrilhes (2)].

Proxim also use HomeRF as a migration technology for devices that originally complied with the OpenAir specification, which no-one was willing to support [Frost & Sullivan].

## 7.2 Architectures

### 7.2.1 System architecture

In HomeRF terminology, stations that act as data sources or recipients are called *nodes.* Particular classes of nodes with identical requirements are formed according to the traffic requirements.

HomeRF networks permit the same network architectures as 802.12. This means there is support both for communication between nodes and communication between a *Central Point*, also sometimes called a *Connection Point* (CP), and several nodes. In addition, HomeRF version 2.0 onward also supports roaming nodes from the cell of one central point to the cell of a different central point.

There are different classes of nodes that reflect the different traffic types listed in section 2.3.1:

- asynchronous nodes (A-Node) support classic data traffic. Applications are, for example, laptops.
- synchronous nodes (S-Node) permit the transmission of constant datastreams (*Streaming Data*), as required, for example, for an audio headset.
- isochronous nodes (I-Node) support bi-directional traffic as found in classic voice telephony.

In addition to this, stations can also involve several traffic types. This permits the convergence of end devices in the sense of Internet appliances. For example, a webpad can receive both classic data traffic and also constant datastreams.

## 7.2.2    Protocol architecture

The HomeRF standard describes the two lowest layers of the reference model (see also Figure 7.1). However, these layers are specially designed to suit the existing protocols of the layers stacked above them. Primarily they assume that three data streams will be used:

- The data path *(Ethernet Data Path)* is designed to suit classic data traffic in digital networks. Most especially, it allows access via the connection-oriented TCP/IP protocol. The data path itself uses a connectionless CSMA/CA access that is very similar to the one defined in IEEE802.12. The datapath is used for connecting asynchronous nodes.
- The media path *(Streaming Media Path)* has been optimised for access via the connectionless and unreliable UDP/IP protocol which especially suits the requirements of music or video data. The media path uses a prioitised CSMA protocol to access the transmission channel at regular intervals. It is used for communications between synchronous nodes.
- The voice path *(Toll-Quality Voice Path)* is designed for isochronous, bi-directional and symmetrical data transmission as defined in the higher protocol layers of the DECT standard. The voice path uses a TDMA process which is inserted into the CSMA accesses. It implements traffic between isochronous nodes.

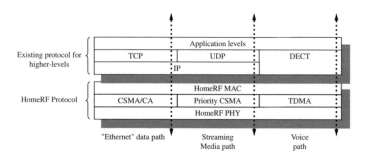

*Figure 7.1:
Structure of the
HomeRF
standard*

## 7.3   Channel access

### 7.3.1   Where it fits in

The protocol implemented in the MAC layer in the HomeRF standard is known as the *Shared Wireless Access Protocol – Cordless Access* (SWAP-CA). Sometimes, the names HomeRF and SWAP are incorrectly used as synonyms. This is especially true in older publications.

The channel is accessed using one of the three service types described below, and the CP manages the access. Figure 7.2 shows the basic processing of a frame. At the start of every frame the PHY makes a frequency hop (see also section 7.4).

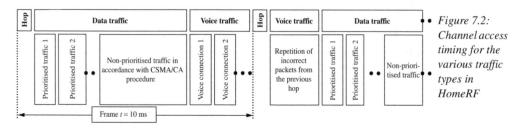

*Figure 7.2:
Channel access
timing for the
various traffic
types in
HomeRF*

Here, in each time period, two main elements are involved in data and voice traffic:

- Data traffic first completes the processing of the prioritised datastream at the beginning of each frame before accepting the normal, unreported traffic from the other data nodes. Here, the prioritised datastreams have priority at the beginning of each frame, even if they require different bit lengths.

- At the end of each frame the voice traffic is processed. Each frame is assigned fixed time slots for the downlink and uplink in a similar way to a TDMA/TDD process. If the voice traffic is not transmitted correctly for any reason, for example due to a disruption of the channel, it can be transmitted again at the beginning of the next frame (re-transmission).

### 7.3.2    Asynchronous transfer

Asynchronous traffic is processed within an extended frame structure, the *superframe*. The duration of each frame is 20 ms (see also Figure 7.3). The access uses a CSMA/CA process which is very similar to the IEEE802.11 standard. Successfully-transmitted packets are confirmed with an acknowledgement of receipt.

*Figure 7.3:*
*Transmitting*
*asynchronous*
*packets over*
*HomeRF*

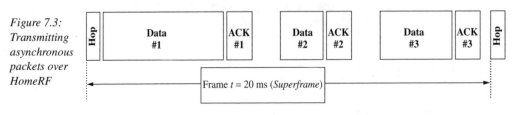

### 7.3.3    Synchronous transfer

The HomeRF 2.0 standard supports up to eight prioritisable bi-directional *Streaming Media Sessions* for audio or video applications.

### 7.3.4    Isochronous transfer

If isochronous traffic needs to be processed for an I-Node, this leads to three changes in Figure 7.4:

- The frame length is shortened to 10 ms. In this way the latency time for the transmission of voice signals can be kept as short as possible.

- At the start of the frame the CP sends a control signal (a *Beacon*) to inform other stations active in its transmission area that two time slots have been reserved at the end of the frame for isochronous traffic.

- At the end of the frame, in a time multiplex (TDD), the voice traffic from the CP to the station (*Downlink*), and from the station to the CP (*Uplink*), is processed. Here two points are worth noting:

  - For every frame, 320 bits need to be transferred in each direction if support is to be provided for voice telephony using ADPCM encoding based on G.726, with a data rate of 32 kbps per direction of transfer, as adopted by DECT. With a bit duration of 625 ns in the HomeRF2.2 standard, which operates at a quick rate of 2.6 Mbps, a partial frame of this kind therefore contains 0.2 ms of control information.

  - The fact that the transfer is first made in the downlink direction, and then in the uplink direction, simplifies synchronisation during channel access, as the downlink transfer is made from the central node.

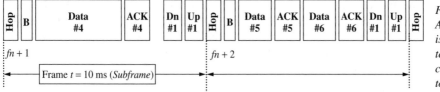

*Figure 7.4: Adding an isochronous telephone conversation to asynchronous data traffic*

If we now add another telephone conversation we see the changes in the first frame with $f_{n+3}$ in Figure 7.5. The *Beacon* control signal now reserves four time slots at the end of the frame for isochronous traffic. First the data from the two stations is transferred in the downlink direction, and then in the uplink direction.

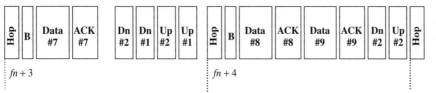

*Figure 7.5: Adding two isochronous telephone conversations to asynchronous data traffic*

If, during the second conversation, the first conversation ends, it is necessary to rearrange the transmissions for the second conversation to provide as long a time interval as possible for the transmission of the asynchronous data. This occurs in the second frame, with $f_{n+4}$ in Figure 7.5.

Once the second conversation ends only asynchronous data traffic still needs to be processed. This means that the control signal is now removed and the frame is once again lengthened to 20 ms.

The frame once again has the same format as shown in Figure 7.3, except that now it is transmitted using the next frequency, $f_{n+5}$.

Version 2.0 of the HomeRF standard supports up to four simultaneous isochronous connections. In the planned version 2.1 of the HomeRF standard it is planned to extend this to eight.

## 7.4    Bit transfer

### 7.4.1    1 MHz channels

In the Physical layer the HomeRF standard uses the Frequency Hopping Spread Spectrum process (FHSS) in the 2.4 GHz ISM band. This is an attempt to match the costs for the Physical layer in Bluetooth modules. The demands made on the components are so similar that it is also possible to create combined radio/transceiver solutions, as described for example for the implementation of Bluetooth, HomeRF and Upbanded DECT in [Vinayak].

Phase shift keying is used for message transfer.

Due to the restrictions in bandwidth caused by the released 2.4 GHz band, 75 base channels are available. Each has a bandwidth of 1 MHz. These base channels are used for data transfer at the speed levels 0.8 Mbps and 2.6 Mbps. They are used to transfer all traffic types and were defined in HomeRF version 2.2.

## 7.4.2    5 MHz channels

If higher data rates are also to be implemented it is necessary to implement the rules for the use of frequency hopping in the 2.4 GHz ISM range defined by the American regulator, the FCC (see also section 1.5):

- The maximum output power must not exceed 1 W.
- At least 75 channels of maximum 1 MHz must be used, and their 20 dB bandwidth must not overlap.
- The average busy time of a frequency, the time between two frequency changes, must be shorter than 0.4 s in a 30 s interval.

This provides a way to increase the speed of transmission provided:

- either the number of bits per transmitted symbol is increased (which unfortunately is also much more likely to mean errors, and is therefore not a feasible solution)
- or the bandwidth of the channels is increased. However, this infringes the rules originally defined by the FCC.

Nevertheless, in a decision made in August 2000, the FCC upheld claims made by the HomeRF working group against resistance from the representatives of WECA and extended the rules on the usage of frequency hopping systems to state that. The maximum output power of 1 W is now reduced to 125 mW and channels can have a maximum bandwidth of 5 MHz and must not overlap at their 20 dB bandwidth. This then provides a total of 15 channels for the released frequency range.

All the same, only data traffic can be transferred at such high speeds. Isochronous data transmission is still carried out using the base channels. Due to the increased bandwidth, only reduced channel density is possible.

# 8 HiperLAN/2

## 8.1 The standard

### 8.1.1 Supplier positioning

The *High Performance Radio Local Area Network, Type 2* (HiperLAN/2) specification was released by ETSI in April 2000, within the framework of the BRAN (*Broadband Radio Access Network*) project. It defined the characteristics of installations for access to fixed networks, in both private and public environments, with bit rates of up to 54 Mbps over distances of up to 150 m, using the 5 GHz range. The main aim was to ensure the provision of services (*Quality of Service* (QoS), and a flexible and extendable service model.

HiperLAN/2 is promoted by the *HiperLAN/2 Global Forum* (H2GF, *www.hiperlan2.com*), which was founded in September 1999 by the companies Bosch (now called Tenovis), Dell, Ericsson, Nokia, Telia and Texas Instruments. By mid-2001, 50 companies had joined this forum.

There has been further activity by the *5 GHz Industrial Advisory group* which, since the start of the year 2000, has had the aim of establishing a world-wide uniform WLAN standard in the 5 GHz range (*http://www.microsoft.com/HWDEV/wireless/5GHz.htm*).

Before the release of the HiperLAN/2 specification the *Radio Equipment & Systems* 10 (RES10) specialist group in ETSI had already defined an initial version of the HiperLAN standard. This was designed for the ad hoc networking of portable devices and is based on a CSMA/CA algorithm. It already supported both asynchronous data transmission and applications with real-time requirements. Once BRAN had been founded the RES10 expert group was dissolved.

However the standard found little support in the industry. Only Proxim's Rangelan 5 system, which it brought out in 1998, was based on HiperLAN/1, but it was quickly withdrawn from sale after a short time.

The original HiperLAN/2 standard has been extended, with two new components offering additional areas of application. The HiperACCESS protocol (formerly "HiperLAN Type III") has a point-to-multipoint architecture, and is designed to offer access for residential areas and business customers for ranges of up to 5 km. The typical planned data rate for HiperACCESS is 27 Mbps for the *Wireless Local Loop* (WLL) . HiperLINK is used for point-to-point links with very high data rates of up to 155 Mbps over distances of up to 150 m. In particular it supports HiperLAN/2 and HiperACCESS interconnections over short routes.

A frequency band of 17 GHz is reserved for HiperLAN Type IV, formerly known as HiperLINK.

Figure 8.1 shows an overview of the various components. You will find summaries of the structure and functionality of HiperLAN/2 in [Walke], [Khun-Jush] and [Johnson]. Here it must be emphasised:

- that HiperACCESS and HiperLINK are completely independent systems which were discussed at the same time, in the BRAN project, in order to make best use of any synergies between them.

- that HiperLINK has more or less been withdrawn from sale for some time now.

*Figure 8.1:*
*Components of*
*the HiperLAN*
*standard*

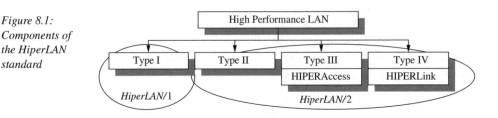

## 8.1.2    Market launch

While, in Europe, ETSI has created the regulatory framework conditions necessary for HiperLAN/2, the following developments can be seen in the USA and in Japan:

- In Japan the *Multimedia Mobile Access Communications* (MMAC) Promotion Association, which is part of the *Association of Radio Industries and Broadcasting* (ARIB), decided to adopt a standard that is fairly similar to the HiperLAN/2 specification. This is called the *High Speed Wireless Access Network* (HiSWANa). In Japan, due to the different regulatory situation, it has also been necessary to implement a mechanism for monitoring the carrier signal (*Carrier Sensing*), as well as using different frequency allocations. In addition there are many more differences, when it comes down to detail [see Aramaki 2001]. In the meantime ETSI, H2GF and the MMAC have formed a joint working group whose aim is to harmonise the two 5 GHz standards. One way to make them mutually compatible would be to make the system elements that are currently required into optional ones.

- In the USA the situation is much more difficult as the IEEE802.11a specification is also positioned in the UNII band. IEEE802.11a and HiperLAN/2 carry out channel access in different ways, but they both use virtually the same OFDM modulation procedure. You will find more information on the prerequisites for the different technologies to succeed in the market place in section 8.6.

# 8.2    Architectures

## 8.2.1    System architecture

A HiperLAN/2 network typically consists of several *Access Points* (APs). They provide full or partial radio coverage in a particular geographical area, called a "cell". In these cells the mobile participants, known as *Mobile Terminals* (MT), communicate with these access points. As well as supporting *Centralized Mode* (CM), in which the mobile participants transfer all useful data over the access points, it also supports *Direct Mode* (DM). In this mode the mobile participants located within transmission range of each other can directly exchange useful data under the control of a *Central Controller* (CC). Support must always be provided for DM, but CM support can be offered as an optional extra, depending on the device profile.

## 8.2.2    Protocol architecture

The specifications of the ETSI/BRAN systems are positioned in the lowest three layers of the ISO/OSI reference model. Figure 8.2 shows the simplified service model of HiperLAN/2. In the Physical layer there is a transmission/security layer which, in HiperLAN terminology, is called the *Data Link Control* (DLC). It is split into three functional units: the *Radio Link Control* (RLC), *Error Control* (EC) and *Medium Access Control* (MAC).

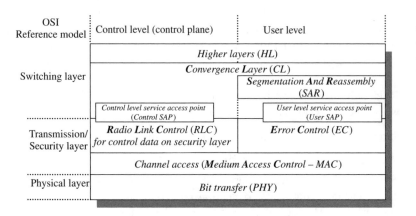

*Figure 8.2: Simplified service model of HiperLAN/2*

To create the transition to the higher layers the *Convergence Layer* (CL) shown in Figure 8.2 has been implemented: it converts the control- and user-level data from the higher layers to the data in the security layer. To achieve this the convergence layer is split in two, once again. A common basis, the *Common Part*, is used to create a service-specific layer, the *Service Specific Part*. In this way you can provide other network services with the usual service access points (SAPs).

This is especially useful for providing access to UMTS core networks for supporting roaming and handover. In this way the supporters of HiperLAN/2 technology hope that mobile phone operators, especially in hot spots, will be able to use HiperLAN/2 to provide broadband access to UMTS as there are clear cost advantages in doing so.

To do so, both the data frames and the control information can be modified as required. An example of how the control information can be modified is that, when HiperLAN/2 is used to access the UMTS network, a handover is triggered on the higher UMTS level, which then also triggers a handover on the HiperLAN/2 level. The CL makes the necessary modification.

*Figure 8.3:*
*Service model*
*of the*
*HiperLAN/2*
*convergence*
*layer*

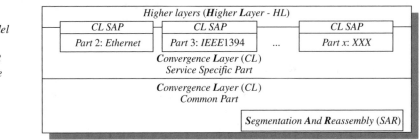

This layer is also responsible for the *Segmentation And Reassembly* (SAR) of *Protocol Data Units* (PDU) on the user side. It is worth noting that HiperLAN/2 is not optimised for a particular network protocol or particular applications. The use of different convergence layers also allows different network protocols such as Ethernet, ATM or IEEE1394 to use the data transfer resources of HiperLAN/2. In this way HiperLAN/2 can also implement connection to the *core networks* that are present.
SAR is only needed for packet-based networks. Integration into cell-based networks such as ATM is not considered here.

With regard to instances in mobile stations, the following applies:

• An RLC instance is generated for every mobile station.

• HiperLAN/2 is a connection-oriented system. It generates an EC instance and SAR instance for each data transfer connection, before actual data transmission. The logical connections must be established before the transfer of useful data. These connections can support point-to-point, point-to-multipoint or broadcast connections.

On the access points side the same number of RLC, EC and SAR instances then need to be generated as the total number generated in the assigned mobile stations.

Using this connection-oriented approach it is possible to support *Quality of Service* (QoS). For each connection, individual QoS parameters can be agreed for data rate, variance, jitter and loss rate.

To do so, the error control protocol (EC) allows the following services:

- in *Acknowledged Mode*, successful reception of a transmission is confirmed with a reply;
- as acknowledged mode is not suitable when data is being sent to several receivers, *Repetition Mode* is used to increase the reliability of the transfer. To achieve this the sections of the packets that contain useful data are sent twice;
- in addition, *Unacknowledged Mode* can be used for simple and fast transfer, but it is only suitable for unreliable connection types.

## 8.3 Channel access

The MAC sublayer is used to control access to the medium: it is managed centrally from the access point using a master/slave protocol approach. The access point assigns capacity to the mobile stations which then use it to transfer useful and control data. When doing so the access point must ensure it takes into consideration the services needed for the various connections.

To do so, the physical channel in the frame is assigned a constant length. A frame length of 2 ms and a symbol time at OFDM level of 4 s correspond to 500 symbols per frame. The frame is then split once again in the transfer phases, with differing functionality, shown in Figure 8.4. The capacity of the individual channels changes as required.

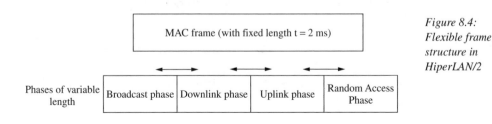

*Figure 8.4: Flexible frame structure in HiperLAN/2*

To understand how channel access works in HiperLAN/2 it is important to note the way that the logical and transport channels are split in the *Data Link Control* (DLC) layer. A logical channel is the path offered by the MAC unit for a particular data transport service. Certain transport channels are used to create logical channels. Transport channels differ in the message formats they use.

Figure 8.5 shows the defined channels and their assignments in a downlink:

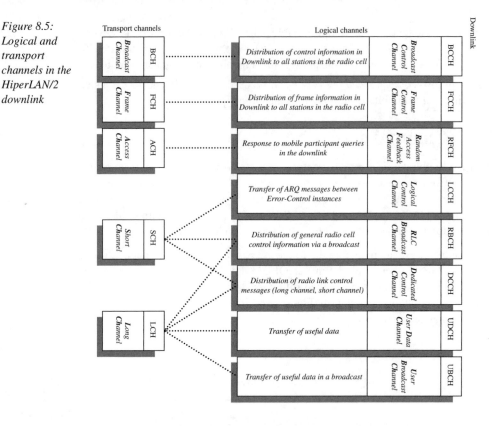

*Figure 8.5: Logical and transport channels in the HiperLAN/2 downlink*

Figure 8.6, on the next page, shows the defined channels and their assignments in an uplink.

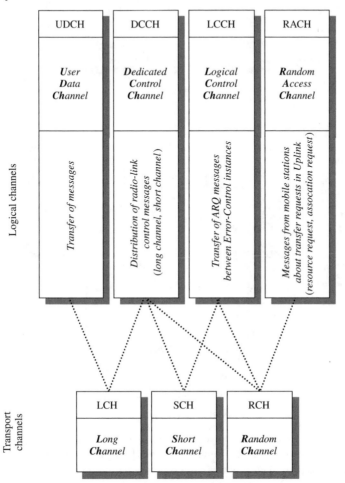

*Figure 8.6:*
*Logical and*
*transport*
*channels in the*
*HiperLAN/2*
*uplink*

# 8.4   Bit transfer

## 8.4.1   Frequency allocations

As section 4.7.1 mentioned, when referring to devices that comply with IEEE802.11a, whose development is primarily being pushed by American companies, the use of the 5 GHz frequency band is also not uniform. The world-wide availability of these channels is shown in Figure 8.7.

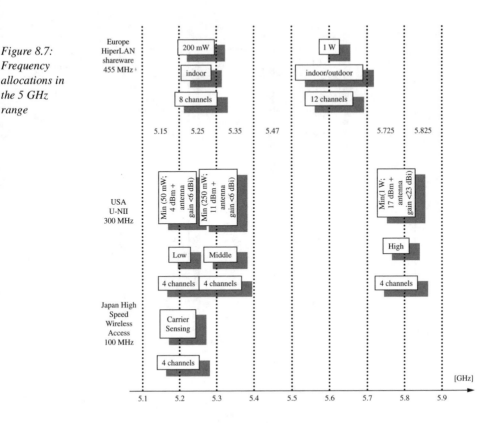

*Figure 8.7:*
*Frequency*
*allocations in*
*the 5 GHz*
*range*

Besides compliance with the frequency ranges, the following additional regional special features need to be taken into account.

- In Europe the regulations also specify the implementation of a regulated transmission output, the *Power Control* (PC), also known as the *Transmit Power Control* (TPC), with a wide-ranging reduction in performance to 3 dB, and *Dynamic Frequency Selection* (DFS), also known as *Dynamic Channel Selection* (DCS). The range from 5.15 to 5.35 GHz is intended for use inside buildings with an average maximum transmitter power of 200 mW (EIRP), while transmissions in the frequencies between 5.47 and 5.725 GHz can have a maximum average output power of 1 W (EIRP). The entire range is available for use without a license (*license exempt*) provided the rules are complied with.

- In the USA neither PC nor DFS are required. Accordingly these functions are not defined in the IEEE802.11a standard which applies in this range. The transmitter powers of 200 mW inside buildings and 1 W outside for the range between 5.15 and 5.35 GHz are maximum values (EIRP). In the range between 5.725 and 5.825 GHz even 4-W transmissions are permitted.

- In Japan only 100 MHz of bandwidth is available between 5.15 and 5.25 GHz. The regulations there require *Carrier Sensing* to occur every 4 ms.

When discussing the problems associated with frequency allocation, regional differences are not the only problem: some applications also already exist in the 5 GHz range. They are mainly satellite and navigation systems. The satellite operators have exceptionally strong political influence.

## 8.4.2 Modulation procedures

HiperLAN/2 uses an *Orthogonal Frequency Division Multiplex* (OFDM) procedure, as described in section 2.8. OFDM achieves high performance, even in dispersive channels such as those that occur in frequencies in the multi-gigahertz range. With it, systems can be implemented that are relatively simple and cheap, compared to the FDM functionality present in HiperLAN/2. Also used is *Multicarrier Modulation*. In it the data is split into different data streams which are then independently transferred over different subcarriers, each with a bandwidth of around 300 kHz. Figure 8.8 shows the 52 subcarriers per channel available. 48 of these subcarriers are used for data transport and 4 as pilot subcarriers for synchronisation. HiperLAN/2 channel therefore has a bandwidth of 20 MHz.

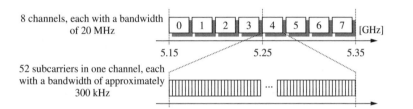

*Figure 8.8: A HiperLAN/2 channel, split into 52 subcarriers*

The different data rates in the Physical layer are then achieved by using the seven modulation types and code rates for *Forward Error Control* (FEC) shown in Table 8.1. Modes 1 to 6 must be implemented in every device, but mode 7, which supports 64QAM, is optional.

| Mode | Modulation | Code rate | Gross data rate | HIPERLAN/2 | IEEE802.11a |
|------|-----------|-----------|-----------------|------------|-------------|
| 1 | BPSK | 1 / 2 | 6 Mbps | Yes | Yes |
| 2 | BPSK | 3 / 4 | 9 Mbps | Yes | Yes |
| 3 | QPSK | 1 / 2 | 12 Mbps | Yes | Yes |
| 4 | QPSK | 3 / 4 | 18 Mbps | Yes | Yes |
| 5 | 16QAM | 1 / 2 | 24 Mbps | No | Yes |
| 5 | 16QAM | 9 / 16 | 27 Mbps | Yes | No |
| 6 | 16QAM | 3 / 4 | 36 Mbps | Yes | Yes |
| 7 | 64QAM | 3 / 4 | 54 Mbps | Yes | Yes |

*Table 8.1: Modulation types in HiperLAN/2 and IEEE802.11a*

Modulation types 1 to 4 are identical in HiperLAN/2 and IEEE802.11a, as the two committees worked closely together to define the Physical layer.

## 8.5   Other services

HiperLAN/2 offers various other services which make it secure and easy to use. Extensive functions ensure high levels of security. It is especially worth noting that these functions, unlike in IEEE802.11, are an integral part of the standard and therefore also function between devices produced by different manufacturers. HiperLAN/2 also uses a number of functions to support ad hoc networking.

## 8.6   The HiperLAN/2 standard versus IEEE802.11a

In the area of 5 GHz networks there are two competing technologies, the HiperLAN/2 standard and IEEE802.11a. Although they are more or less identical at the level of physical transfer they differ a great deal at the level of channel access. From the technical point of view the HiperLAN/2 standard clearly seems superior, as the overview below, in Table 8.2, shows. However, to succeed in the marketplace, other political and company-related aspects, and cost issues, are just as important.

*Table 8.2: The HiperLAN/2 standard compared with IEEE802.11a*

| Feature | IEEE802.11a | HiperLAN/2 |
|---|---|---|
| Dynamic frequency selection | No | Yes |
| Adjustment of transmitter power | No | Yes |
| Security | Partially | Yes |
| Support for QoS | No | Yes |
| Integration with other standards (UMTS) | No | Optional |

# 9 Operating an IEEE802.11b-compliant WLAN

## 9.1 Introduction

This chapter provides an insight into the operation, administration and analysis of wireless local networks (WLANs) that comply with the IEEE802.11b standard. Various manufacturers have kindly provided equipment that the author has used to describe such a network. These series products provide a good insight into typical implementations and show the advantages and disadvantages of each implementation.

## 9.2 Mobile stations

### 9.2.1 Construction types

Mobile stations have a variety of construction types for PC-based applications, as described in section 3.2.

### 9.2.2 Capabilities

Some fundamental properties have a major influence on how user-friendly the equipment is. These criteria include the following:

- During installation you hardly notice any differences. Most systems allow their drivers to be installed using the MS Windows Hardware Manager, an easy-to-use wizard. However some devices turned out to be sensitive when several different PC cards were in operation, requiring the author to restore his previous operating system installation during the tests he made.

- All systems support the operating systems of the Microsoft Windows family. The commercial standard packages generally did not support Unix, and in particular did not support Linux. For use under Linux, the currently-available cards are integrated in the Linux kernel from kernel version 2.2. For other products you have to rely on the willingness of other Linux users to help [Tourrilhes (1)].

- Almost all systems contain menu-driven tools which allow simple administration of the network connection and the analysis of the networks present. Generally these need to be installed separately from the hardware drivers. A wizard is usually provided for this purpose.

- As the systems supplied need to be easy to use, and in particular must also meet the security requirements of other networks, dedicated analysis tools are provided. These are described in more detail in section 9.5.

### 9.2.3  Administration

The mobile stations are managed using menu-driven graphical user interfaces. The ease with which the devices from different manufacturers can be set up, and the number of settings that can be made for each device, vary considerably.

*Figure 9.1:*
*Aironet Client*
*Utility from*
*Cisco –*
*network*
*architecture*
*parameters*

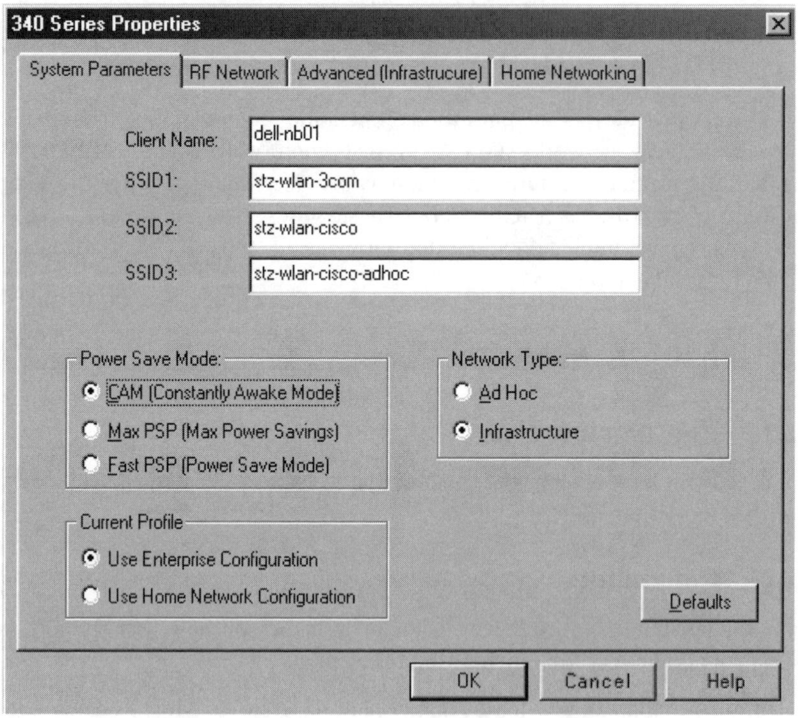

Figure 9.1 shows the Parameters tab in the Cisco Aironet Client Utility, in which the most basic properties of the network architecture can be set. Here the most important question is whether infrastructure mode or ad hoc mode is required. You also specify the permitted SSIDs and the main energy management settings. Here you should note that not all client products support ad hoc mode. You can overcome the security barrier of SSID assignment on many products by specifying the ID *any*, which gets past the SSID check.

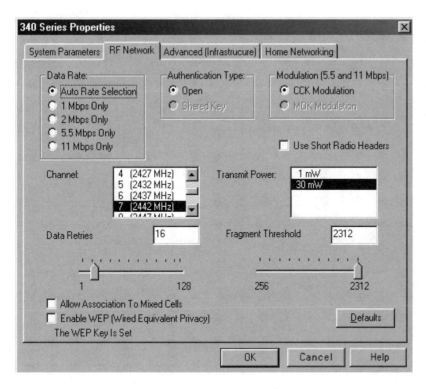

*Figure 9.2:
Aironet Client
Utility from
Cisco –
physical
transfer
parameters*

Using the next menu, shown in Figure 9.2, you can set the basic parameters for physical transfer. These include, among others:

*   selection of the speed levels (data rates) and modulation types. Here, Cisco Aironet users can also still select the old MOK standard modulation type (see also section 4.7.2) to ensure backwards compatibility with older systems that only support MOK.
*   selection of the channel. Here you should remember that this parameter, unlike in other systems such as DECT (see Chapter 6), or HiperLAN/2 (see Chapter 8), is entered statically by the user. For the mobile station this only affects the sequence in which it is to monitor the channels. In Infrastructure mode the access point specifies channel utilisation (see also section 9.3.3).
*   setting of the transmitter power. The most usual setting will be 30 mW so that a 10 dBi antenna can still be used. However, the transmitter power can also be reduced. This would be appropriate if antennae with large gain levels are to be used, or for operating power-saving systems over short distances.
*   activation of encryption (see also sections 4.6 and 10.1).

In the tab shown in Figure 9.3 you can make additional settings for integrating the device in a network. These settings especially include the *Antenna Diversity*, the assignment of permitted access points and the setting of transmission parameters (see also section 2.9.2).

*Figure 9.3:*
*Aironet Client*
*Utility from*
*Cisco –*
*additional*
*network*
*architecture*
*parameters*

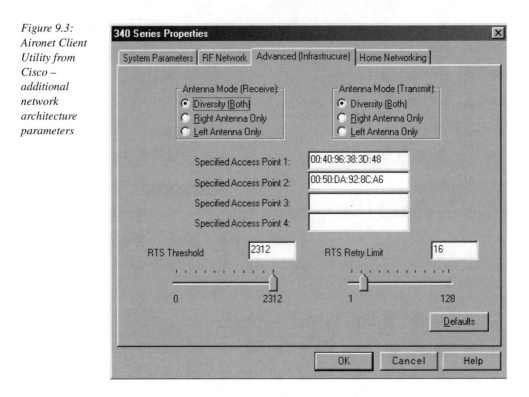

The final tab in the Cisco Administration tool, Home Networking, has an interesting, but currently rather half-hearted, purpose. Besides the settings discussed above, for setting up the device in a more complex office environment, a second default configuration, containing considerably fewer parameters, is also possible. It is aimed at the user who, besides using their laptop in the office, also wants to connect to the network at home, via a radio LAN. You can easily imagine that future versions could contain the main settings for several network environments, allowing the user to move between them conveniently.

Finally, most devices are also supplied with additional analysis tools. They are discussed in more detail in 9.5.2.

# 9.3    Access points

## 9.3.1    Structure

You install *Access Points* (AP) in a different way from the procedure for bringing client-side adapter cards into operation. Access points are network nodes that work autonomously and which, under certain circumstances, can take on other infrastructure tasks. Access points usually have their own embedded microprocessor for controlling processes.

There are two ways to arrange communication and access points administration:

*   If the network connection is used for communication, an embedded HTTP server is usually accessed on the access point's microprocessor, where it is active. This permits access using an HTTP client, which is a normal web browser such as Microsoft Internet Explorer or Netscape Communicator. For this purpose the webserver needs its own fixed IP address. However, this IP address is only used for addressing the webserver, not for addressing the net node like a router.

    The network connection can take two forms:

    *   via the wired network to which the access point creates the connection, or
    *   via the wireless network. This second alternative is the only one possible if an autonomous wireless infrastructure network is operating. However you should then note that the access point itself is managed using the access that is to be managed.

*   All access points also have a serial port. This is needed especially for ensuring that the access point can be accessed for its initial installation. The serial port can be addressed easily by connecting to it with a serial cable from the serial port on a PC, taking into account the fixed parameters that have been set. Figure 9.4 shows the user interface for a typical terminal access.

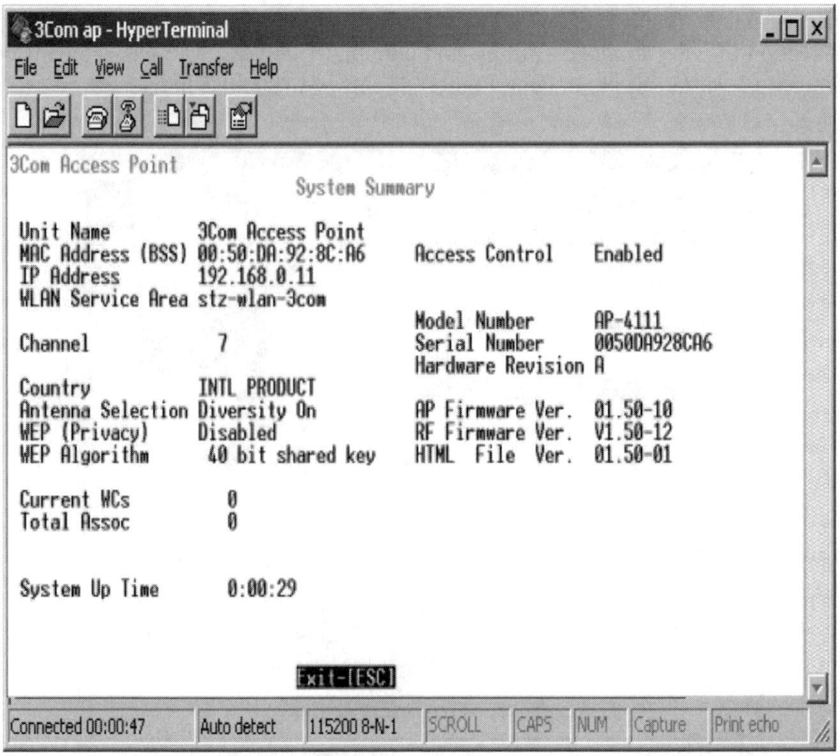

*Figure 9.4: User interface for administering the 3COM Access Point via a terminal program*

### 9.3.2    Capabilities

Access points also differ considerably from each other. For this reason a few criteria are also listed below:

- The methods for accessing the access point described in section 9.3.1 mean that there is no need to install additional administration tools. The only programs needed are a terminal program and web browser, which are standard functionality on PCs.

- A serial cable is required for access via the serial port. Some systems use a null modem cable (cross-over cable), and others use a standard serial cable, so it is very convenient for the user if a suitable cable is supplied with the device.

- The menu-driven utilities on some systems take a lot of getting used to, especially the keyboard-driven user interfaces used in terminal connection programs, which are tricky to use. Even some web-based menus are not always easy to use.

- If you change parameter settings the webserver needs to pass your changes to the access point. When an update is made in this way, some devices reboot, which means that repeated updates take a very long time.

### 9.3.3    Administration

It takes much longer to administer access points than client adapter cards, as access points require many more parameter settings. Below you will find a selection of ways to make settings, and administer devices. The management of logged-on stations is paramount. Figure 9.5 shows all stations in the network.

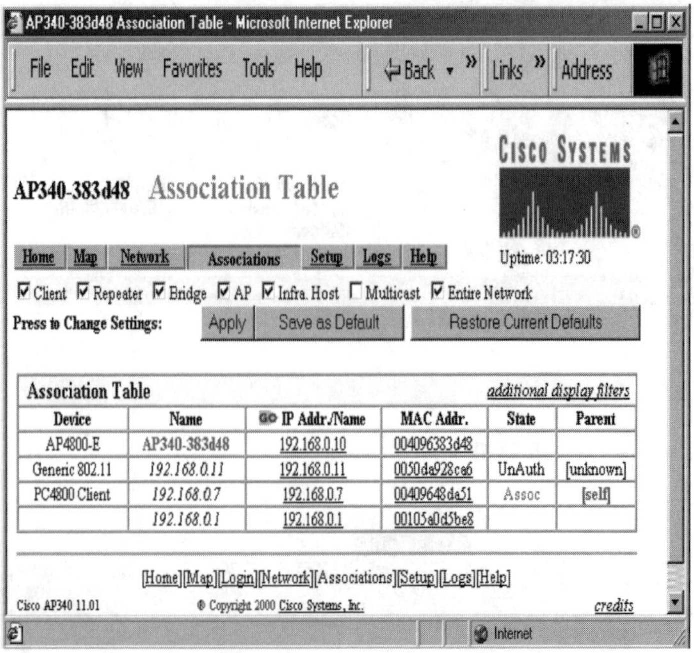

*Figure 9.5:*
*Aironet AP*
*Management*
*from Cisco:*
*association*
*table*

AP340-383d48 Association Table - Microsoft Internet Explorer

File   Edit   View   Favorites   Tools   Help        Back ▼ » | Links » | Address

CISCO SYSTEMS

**AP340-383d48**   Association Table

Home | Map | Network | Associations | Setup | Logs | Help        Uptime: 03:17:30

☑ Client  ☑ Repeater  ☑ Bridge  ☑ AP  ☑ Infra. Host  ☐ Multicast  ☑ Entire Network
**Press to Change Settings:**   Apply | Save as Default | Restore Current Defaults

**Association Table**                                    *additional display filters*

| Device | Name | IP Addr./Name | MAC Addr. | State | Parent |
|---|---|---|---|---|---|
| AP4800-E | AP340-383d48 | 192.168.0.10 | 004096383d48 | | |
| Generic 802.11 | 192.168.0.11 | 192.168.0.11 | 0050da928ca6 | UnAuth | [unknown] |
| PC4800 Client | 192.168.0.7 | 192.168.0.7 | 00409648da51 | Assoc | [self] |
| | 192.168.0.1 | 192.168.0.1 | 00105a0d5be8 | | |

[Home][Map][Login][Network][Associations][Setup][Logs][Help]
Cisco AP340 11.01              © Copyright 2000 Cisco Systems, Inc.              *credits*

Internet

These stations include:

- access point AP4800-E with the IP address 192.168.0.10
- a mobile terminal PC4800 client with the IP address 192.168.0.7. The access point is accessed using this client, as the ID *self* shows.
- a desktop PC acting as a server with the IP address 192.168.0.1. This also takes on the tasks of a DHCP server, SMB server and HTTP server.
- another access point with the IP address 192.168.0.11, whose ID is only shown as *Generic 802.11*. This is another access point made by a different manufacturer (3Com) which is connected over the network.

You can also observe the transport characteristics of the radio route, as Figure 9.6 shows. Here it makes sense to strictly distinguish between the individual network layers. *Ethernet* stands for the MAC layer, and *AP-Radio* stands for the Physical layer.

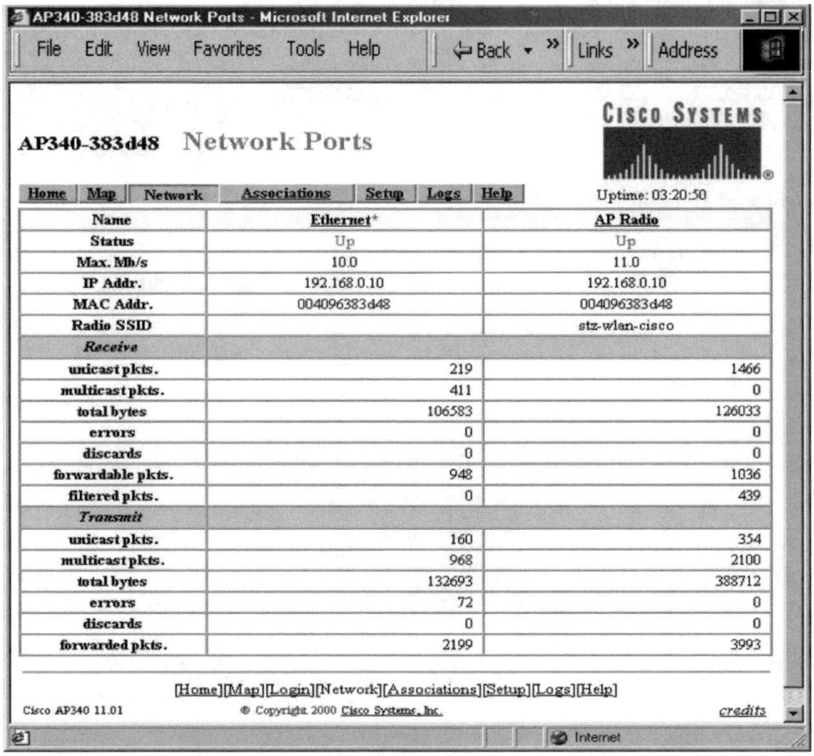


| Name | Ethernet* | AP Radio |
|---|---|---|
| Status | Up | Up |
| Max. Mb/s | 10.0 | 11.0 |
| IP Addr. | 192.168.0.10 | 192.168.0.10 |
| MAC Addr. | 004096383d48 | 004096383d48 |
| Radio SSID | | stz-wlan-cisco |
| *Receive* | | |
| unicast pkts. | 219 | 1466 |
| multicast pkts. | 411 | 0 |
| total bytes | 106583 | 126033 |
| errors | 0 | 0 |
| discards | 0 | 0 |
| forwardable pkts. | 948 | 1036 |
| filtered pkts. | 0 | 439 |
| *Transmit* | | |
| unicast pkts. | 160 | 354 |
| multicast pkts. | 968 | 2100 |
| total bytes | 132693 | 388712 |
| errors | 72 | 0 |
| discards | 0 | 0 |
| forwarded pkts. | 2199 | 3993 |

*Figure 9.6: Aironet AP Management from Cisco: radio route transport characteristics*

Access can be limited to different access levels. Each access point supports an SSID which accordingly must also be entered in the client (see also section 4.6). You can enter the MAC addresses of the authorised stations. This reveals an important security aspect (and the problem it causes):

- as long as the addresses of stations with authentication are known there is no problem with entering details in the access points list. However, as soon as an open group has access to the radio LAN, as is the case in the example in section 9.6.3, it is not possible to make these entries. A typical solution is to extend the limitation to *ranges*. Although this is more convenient it means a reduction in security.

- if several access points are in use, each access point must contain all the MAC addresses. Management either then becomes very time-consuming and error-prone, or a RADIUS server may need to be used.

An important element in monitoring access points is the logging of any logons that occur, and analysing them using automated tools whenever possible, to prevent access by unauthorised participants. Figure 9.7 contains an extract from a log file of this kind, which is stored on the access point.

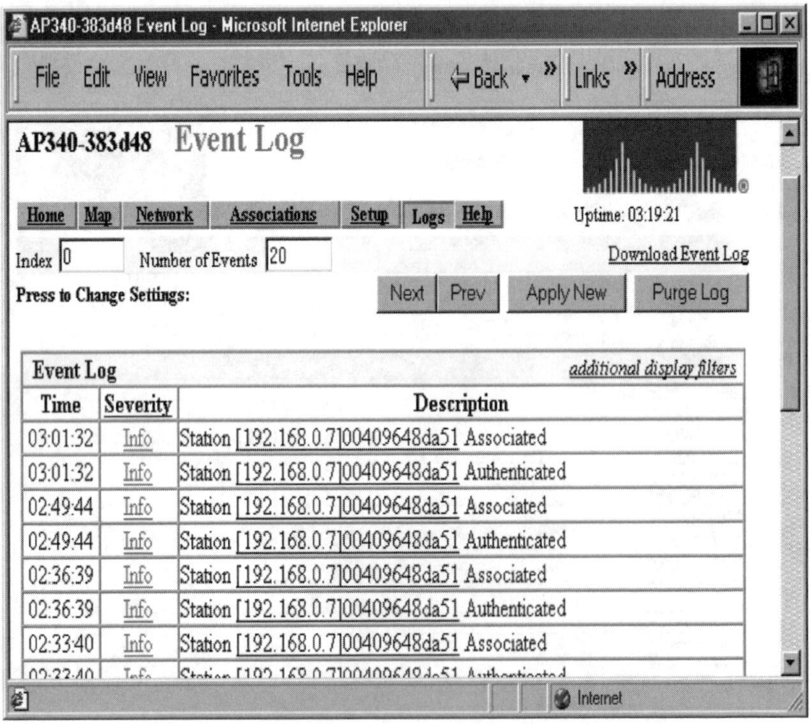

*Figure 9.7: Aironet AP Management from Cisco: log showing all events*

Although the IEEE802.11 standard does not specify **Dynamic Frequency Selection** there are proprietary solutions for finding the channel with the lowest activity level during booting. It is not practical to completely change channel during operation, especially if stations supplied by different manufacturers are in use on the network.

# 9.4    Extended networks

## 9.4.1    Linked networks

When coverage is being provided in large buildings or areas it often becomes clear that several radio stations are needed due to the high levels of attenuation caused by obstacles, restricting the size of the radio cell of an individual access point. It is necessary to set to a linked network, consisting of an *Extended Service Set* (ESS), see also section 4.2.1.

It is relatively simple to set up a network of this kind, especially if it only contains devices of one type. If different devices are used, or even devices supplied by different manufacturers, this can sometimes lead to incompatibilities when it comes to administration.

Figure 9.8 shows the resulting schematic structure. The three access points AP1 to AP3 are physically distributed in such a way that they partially overlap. The aim is for a mobile user, who is moving about the building, to be able to access the network in all areas without changing their configuration. For this purpose:

• all access points use the same channel.

• all access points are assigned the same SSID. It is also possible to assign the access points different SSIDs, but then the mobile client needs to be informed of them (see also Figure 9.1).

• all access points are assigned the same WEP key.

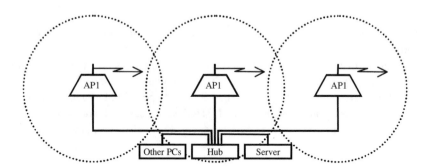

*Figure 9.8: Operating several access points in non-overlapping or partially-overlapping radio cells*

## 9.4.2    Structured networks

If a physical area is to contain different work groups that belong to different domains, the following configuration can be used:

• the access points, and therefore the subnets, work on different channels. If DSSS systems that comply with 802.11b are used, the three overlapping channels 1, 6 and 11 are available.

• the access points are assigned different SSIDs.

• the access points are assigned different WEP keys.

### 9.4.3   High user densities

If areas with high user densities, known as *hot spots*, are to be connected to radio networks, the limited bandwidth needs to be taken into consideration. Typical applications are conference and training rooms, waiting rooms and trade fair halls.

Estimates of the maximum number of users who will work via an access point with a gross data rate of 11 Mbps and a maximum net data rate range of 5 to 6 Mbps largely depend on the ways users behave. If there are a very large number of users, this means that several access points that overlap or cover the same area need to be implemented in parallel (see also Figure 9.9).

*Figure 9.9: Operating several access points with radio cells that overlap or cover the same area*

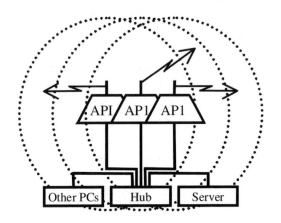

Many currently-available access points support *Load Balancing* for such applications, in which mobile stations are assigned to access points according to their load. However the signalling used for load balancing is not standardised, but manufacturer-specific, For negotiating the assignment both unused control fields in the frame structure (see also section 4.3.6) and special data packets are used. These data packets are sent to special MAC addresses that are otherwise not present in the network. The data packets then also contain the control information.

### 9.4.4   Administering extended networks

The administration of extended networks takes a lot of effort as the configurations of all stations present must be consistent with each other. It is useful to implement a *Remote Authentication Dial-In User Service* (RADIUS) server to administer a network of this kind. These servers are most often sold as *Access Control Servers* (ACS) or as *Access Point Controllers* (APC) by the manufacturers who supply the professional market.

The only problem is that this administration system has some proprietary components.

# 9.5    Network analysis

## 9.5.1    Overview

When bringing a network into operation, or operating it, it is sensible and necessary to take into account a large number of parameters to ensure that the network provides the best possible performance. When administering radio networks there are two areas of interest:

- the parameters for the Physical layer
- the parameters that control the amount of traffic in the logical traffic layers.

There are three main ways to analyse networks in real life:

- the powerful WLAN stations currently available include not only administration tools but also software for analysing any network connections present.

- special analysis tools are available which use the physical details of standard hardware to obtain an overview of the entire wireless radio network at a specific location. These tools are generally called *sniffers*.

- finally, dedicated hardware is available that can be used to determine the characteristics of radio transmission.

## 9.5.2    Simple analysis tools

The best way to test the capacity of a network connection is to transfer large files (see also Figure 9.10). However, you should remember that the speed of a process of this kind only indicates the capability of the entire computer system. For example, if there is insufficient bandwidth, it is not possible to tell if the problem lies in an overloaded network connection, or an overloaded transmitting or receiving computer.

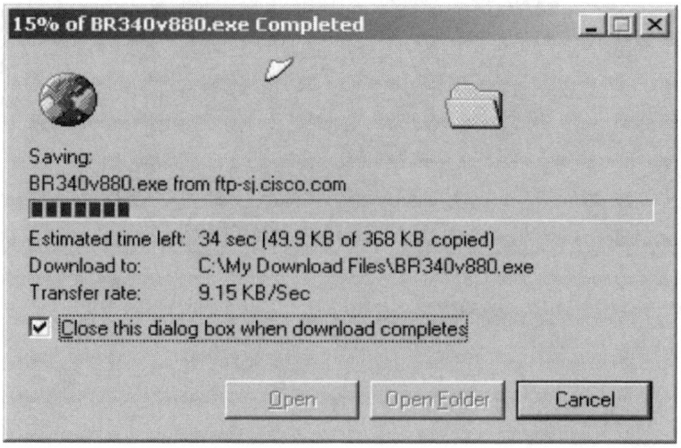

*Figure 9.10:*
*Observing the*
*bandwidth at*
*application*
*level*

This means that it is necessary to analyse these areas separately. Below you will find some typical examples of the capabilities of the tools supplied with WLAN stations, which can be used to analyse radio connections.

Figure 9.11 shows the *Cisco Link Status Meter* with which you can carry out large-scale measurements in a physical space.

*Figure 9.11:*
*Information*
*about the*
*characteristics*
*of the radio*
*field*

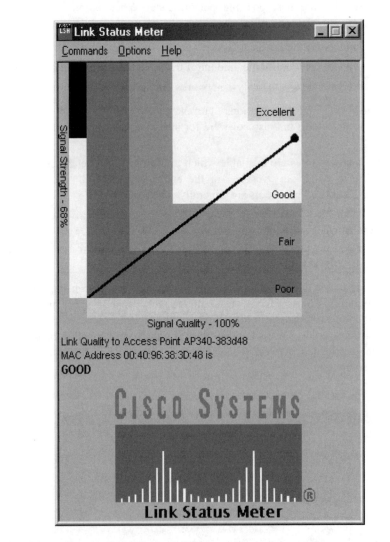

The *3Com Connection Monitor* displays other analyses that cover incorrect transfers, speed levels etc. in Figure 9.12.

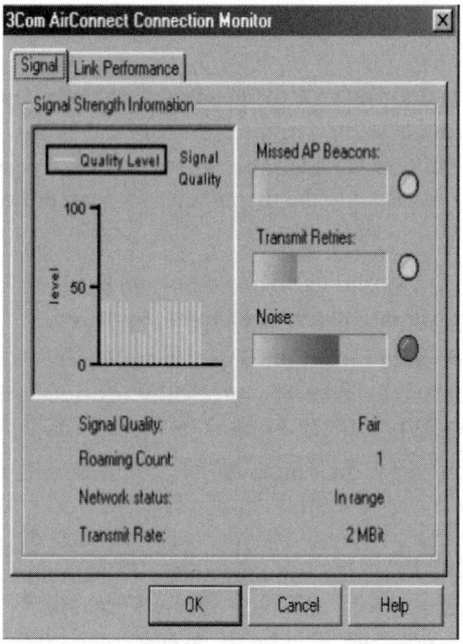

Figure 9.12:
Extra
information
about the error
rate and
transfer speed

Almost always, a tool is also provided in which the ICMP *ping* command is used to test whether stations can be contacted and to measure the propagation delay time (*round trip time*) of the connections. Some tools allow you to set different parameters for this transfer, especially the packet size. Some also provide graphical evaluations (see also Figure 9.13).

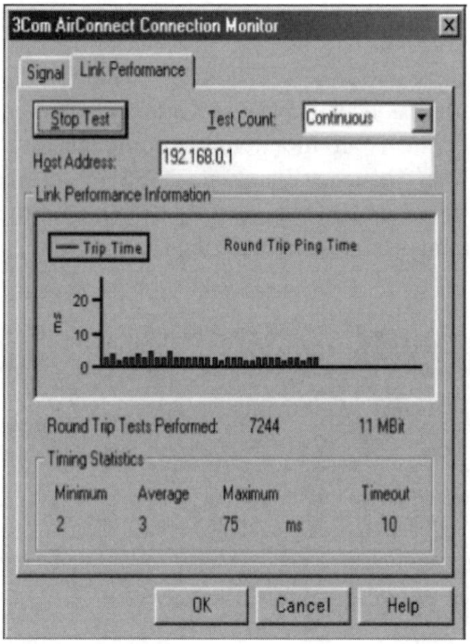

Figure 9.13:
Integration of
the ping
command in
the network
analysis tool

### 9.5.3   Sniffers

**Basics**

A sniffer is a special analysis tool for observing network traffic. The name was coined for a product made by the company *Network General*, but is now used as a generic label. Many sniffers are based on normal network adapters, but they often use special operating modes that the manufacturers have not released for all users. Sniffers are primarily used in two ways:

- to analyse a network to find its weak points and bottlenecks. To do this, traffic streams and physical parameters need to be displayed.

- to monitor a network to obtain information without permission (illegally). As the electro-magnetic fields are not physically restricted, this kind of criminal activity is an especially critical problem in wireless networks.

Some examples below show the functionality provided by a professional analysis tool of this kind. The WLAN Sniffer supplied by Networks Associates Inc. (NAI, *http://www.nai.com* and *http//www.sniffer.com*) has been developed for use with the Cisco Aironet 340-PC Card, among others. During sniffing it does not take part in active communication, and only receives. A special driver needs to be installed before it is possible to directly access other internal registers in the network adapter, which have otherwise not been released by Cisco.

When observing traffic on the radio channel you must take into account the local positioning of the receiving antenna, due to the physical distribution of the electro-magnetic waves. It is especially useful to install two WLAN adapter cards in a laptop, one for processing normal traffic and the second as a sniffer. However, due to some antenna constructions this is not possible for all adapter cards.

**Channel Surfing**

To perform these analyses on all channels a *Channel Surfing* function is provided. With it the various channels are monitored at particular time intervals. You will quickly notice that the frequency bands in DSSS mode overlap so much that packets can also be received and successfully decoded on the active channels in adjacent frequency ranges (see also Figure 9.14).

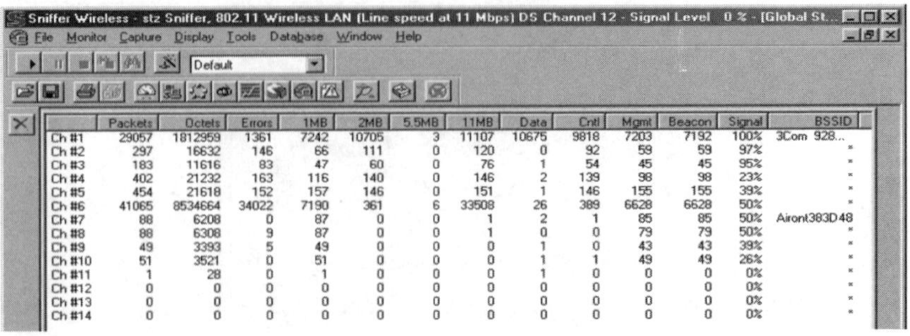

*Figure 9.14:*
*Activities of the*
*various radio*
*channels*

| | Packets | Octels | Errors | 1MB | 2MB | 5.5MB | 11MB | Data | Cntl | Mgmt | Beacon | Signal | BSSID |
|---|---|---|---|---|---|---|---|---|---|---|---|---|---|
| Ch #1 | 29057 | 1812959 | 1361 | 7242 | 10705 | 3 | 11107 | 10675 | 9818 | 7203 | 7192 | 100% | 3Com 928... |
| Ch #2 | 297 | 16632 | 146 | 66 | 111 | 0 | 120 | 0 | 92 | 59 | 59 | 97% | " |
| Ch #3 | 183 | 11616 | 83 | 47 | 60 | 0 | 76 | 1 | 54 | 45 | 45 | 95% | " |
| Ch #4 | 402 | 21232 | 163 | 116 | 140 | 0 | 146 | 2 | 139 | 98 | 98 | 23% | " |
| Ch #5 | 454 | 21618 | 152 | 157 | 146 | 0 | 151 | 1 | 146 | 155 | 155 | 39% | " |
| Ch #6 | 41065 | 8534664 | 34022 | 7190 | 361 | 6 | 33508 | 26 | 389 | 6628 | 6628 | 50% | " |
| Ch #7 | 88 | 6208 | 0 | 87 | 0 | 0 | 1 | 2 | 1 | 85 | 85 | 50% | Airont383D48 |
| Ch #8 | 88 | 6308 | 9 | 87 | 0 | 0 | 1 | 0 | 0 | 79 | 79 | 50% | " |
| Ch #9 | 49 | 3393 | 5 | 49 | 0 | 0 | 0 | 1 | 0 | 43 | 43 | 39% | " |
| Ch #10 | 51 | 3521 | 0 | 51 | 0 | 0 | 0 | 1 | 1 | 49 | 49 | 26% | " |
| Ch #11 | 1 | 28 | 0 | 1 | 0 | 0 | 0 | 0 | 0 | 0 | 0 | 0% | " |
| Ch #12 | 0 | 0 | 0 | 0 | 0 | 0 | 0 | 0 | 0 | 0 | 0 | 0% | " |
| Ch #13 | 0 | 0 | 0 | 0 | 0 | 0 | 0 | 0 | 0 | 0 | 0 | 0% | " |
| Ch #14 | 0 | 0 | 0 | 0 | 0 | 0 | 0 | 0 | 0 | 0 | 0 | 0% | " |

## Statistics

In addition, very detailed analyses of a wireless network's traffic and errors can be produced. The sniffer provides the module shown in Figure 9.15 for that purpose. Obviously, a high load on a channel will lead to very high error rates. This is especially critical because it means that the time slots in which collisions occur cannot be used to transfer useful data. When the load is high, the usable channel capacity drops, as is also the case for wired Ethernet.

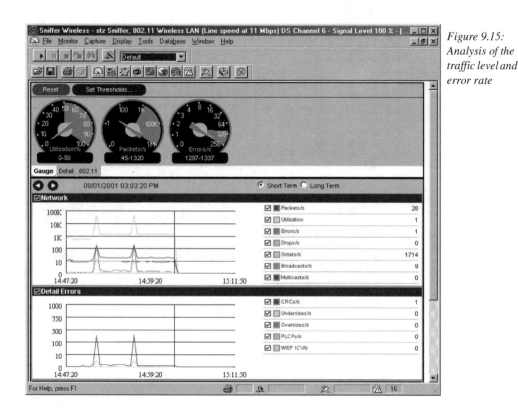

*Figure 9.15: Analysis of the traffic level and error rate*

## Individual analyses

If specific errors need to be analysed, the global analyses shown up to now provide little help. Usually, a detailed investigation of the individual activities on the radio channel is required. Two examples illustrate this:

- Figure 9.16 shows a station logging on to an access point, as described schematically in section 4.3.7.

- Figure 9.17 contains an extract from the traffic in a badly-overloaded network. In this example a file for a website is being transferred. You can see very clearly that two normal transfer attempts (nos. 72 and 73) have been unsuccessful, and that two more attempts are made with the help of the *Clear-To-Send* (CTS) and *Request-to-Send* (RTS) sequence of RTS-CTS signals.

You can see the sequence of RTS-CTS data packets in steps 74 to 76 and 77 to 79. Once these attempts have been made, unsuccessfully, another normal transfer attempt is made (step 80), and confirmed with an ACK in step 81. In the result we find that the successful transfer of 411 bytes required a time interval of 8 ms. This corresponds to a data rate of 411 bps.

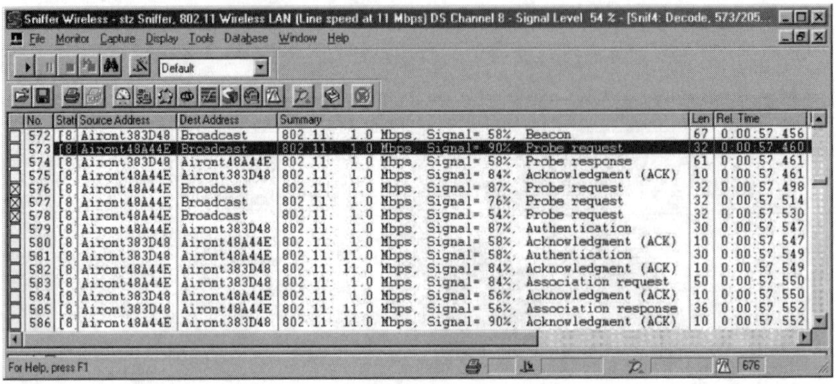

*Figure 9.16:*
*Trace of a*
*station's logon*
*to an access*
*point*

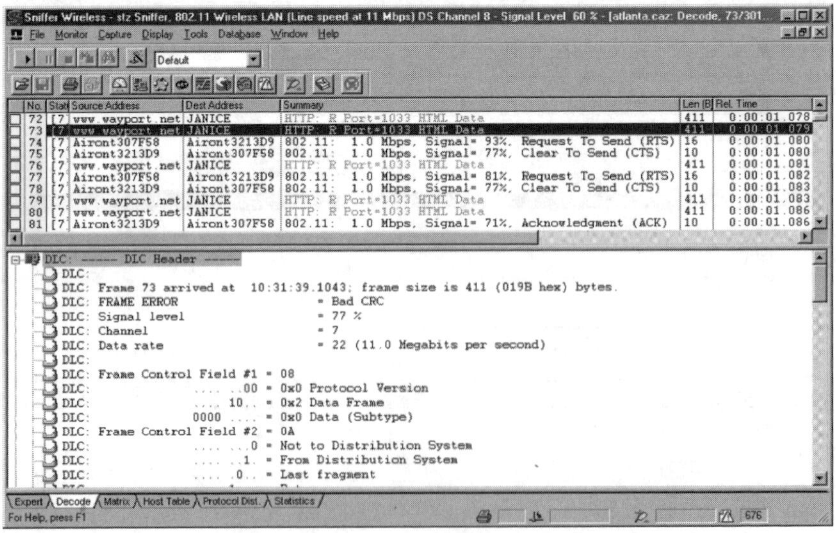

*Figure 9.17:*
*Trace of a*
*multiple*
*transfer*
*attempt using*
*DCF and PCF*

# 9.6    Examples from real life

## 9.6.1    LAN-LAN connection

### Initial situation

The University of Cooperative Education (Berufsakademie Lörrach), in Lörrach, Germany (*http://www.ba-loerrach.de*) is a technical college with around 1000 students who study to become graduates in business and economics, commercial IT or engineering. The actual campus, which includes the central administration department, computing centre and laboratories, is located in several separate buildings on the edge of town. Student numbers rose sharply in the 1990s so another building was rented in the town in 1999.

Here, the following factors come into play:

- Due to the geographical situation, not only offices, but also teaching rooms and a computer room need to be set up there.

- However, there is no line of sight between the campus and this annex building.

- The administrative and teaching staff share one of two separate networks, and students use the other, which are operated to ensure there is sufficient separation between them. These two networks also need to be available at the new location. Due to the number of student accounts required, a total IP address space of three C-classes is available.

## Implementation

The implementation shown in Figure 9.18 was installed in late 1999.

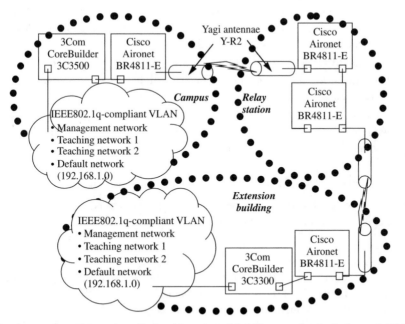

Figure 9.18:
Radio bridge
for creating a
LAN-LAN
connection
between two
subnets

The three networks are bundled over a virtual LAN on a physical channel. This virtual LAN is transported to the annex building using an IEEE802.11b-compatible radio route. The set-up includes an additional relay station on a high public building in the town. This means that a total of four active elements and antennae are required.

**Possible ways to extend the network**

*Figure 9.19:*
*Traffic*
*characteristics*
*of the radio*
*bridge*
*(two-hourly*
*samples)*

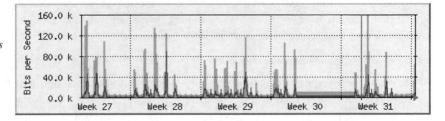

The usage characteristics of the two-hourly samples shown in Figure 9.19 are for a typical time period. The largest amount of traffic in it was created by users in a computer room with 28 workstations.

When traffic loads peak the radio bridge becomes a bottleneck so it is worth considering possible ways to extend the network. In this situation the following four approaches are possible:

- Upgrade to a radio module in the 2.4 GHz range and a data rate of 22 Mbps. In the case of point-to-point-systems such as the radio bridge shown here non-conformity to a future IEEE802.11g standard is not a fundamental problem.

- The use of a radio bridge that complies with the HiperLAN/2 standard, in the 5 GHz range and with a gross data rate of 54 Mbps. However, commercial products will not be available until mid-2002.

- Upgrade to a radio link. Although this can achieve significantly higher bandwidths, it requires much more investment and also needs approval from the relevant authorities.

- The use of up to three parallel radio routes. IEEE802.11b provides three non-overlapping frequency ranges, which can be fully utilised through *Load Balancing* in the third network level.

## 9.6.2    Mobile computer room

A second installation has also been set up in and around the University of Cooperative Education Lörrach, with a typical application which is described below.

IT e-network administrators are trained as part of a further education project, carried out together with IHK (Chamber of Commerce and Industry) Oberrhein (on Lake Constance) and the Steinbeis Transfer Center for Business Consulting at the University of Cooperative Education Lörrach. A mobile computer room has been created for this training course. It is equipped with ten desktop mini-PCs and can be set up in any training room that has access to a wired network. The network connection has been implemented using an IEEE802.11b-compliant WLAN network and one access point.

### 9.6.3   Changing stations in the hotel

The Hotel am Rathaus in Freiburg (*http://www.am-rathaus.de*) is a family-owned hotel in the centre of the old town of Freiburg, a picturesque tourist destination near the Black Forest. The hotel offers its guests not only Internet access, but also many other up-to-date electronic communications services.

They include another service that is available in numerous modern hotels and conference centres: a network connection for the guests via an IEEE802.11-bis-compliant WLAN network. Guests can either borrow a complete laptop or simply a WLAN card that they can then install in their own laptop.

The physical layout of the hotel make it possible to cover an entire floor with one access point, which results in the very simple configuration shown in Figure 9.20. An IEEE802.11-bis system is also used to connect the entire hotel network to its ISP (Internet Service Provider) by means of a point-to-point radio bridge. A router is used to separate the traffic.

In the hotel itself, four stories and the ground floor, which includes the reception area, need to be covered. In this case the radio network also covers the café area in the hotel entrance, providing guests with Internet access in the open air. At present there are no plans to offer Internet access to café customers who are not actually staying at the hotel. Network recognition is used to restrict access.

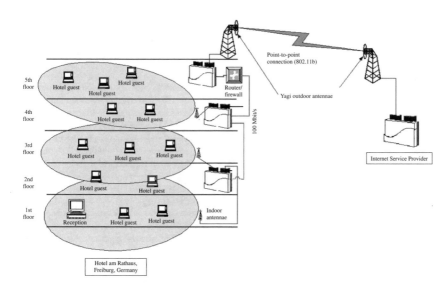

*Figure 9.20: Architecture of network provision in the Hotel am Rathaus, Freiburg, Germany (with the kind permission of Netlight GmbH)*

# 10 Various aspects of WLAN technology

## 10.1 Security

### 10.1.1 Basics

This section describes a range of security issues concerning WLAN protocols which are currently the topic of detailed and lively discussions. It first describes both the basic technology used to achieve security, and the possible associated risks, and then suggests a possible evaluation. The starting point of the discussion is the *Wired Equivalent Privacy* (WEP) protocol, implemented as part of IEEE802.11, as described in section 10.1.2, Security risks.

The current situation can be described in the following way:

- The WLAN systems available in today's market usually have security mechanisms.
- These security mechanisms sometimes have gaps through which attacks could take place. Fundamentally there are two kinds of attack: passive attacks (eavesdropping) and active attacks (penetration).
- Systems implemented in accordance with IEEE802.11 are particularly prone to gaps in security, which is why the discussion below uses them as an example. At present the comparable risks in other WLAN systems have not yet been made public.
- In principle, these activities are also possible with commonly-used mass produced devices. In addition the World Wide Web has freely-available shareware software tools that can easily be used to measure radio fields. At the same time these tools can also be used to discover the most basic information about unprotected networks. Figure 10.1 shows the interface of one of these tools (www.netstumbler).
- In real life we see again and again that these identified risks are ignored and that even existing security mechanisms are not used effectively, making it even easier for attacks to take place. No-one has yet disagreed with this assessment!

What springs to mind in this context is the image of a front door that is protected against burglars.

*Figure 10.1:*
*NetStumbler*
*interface*

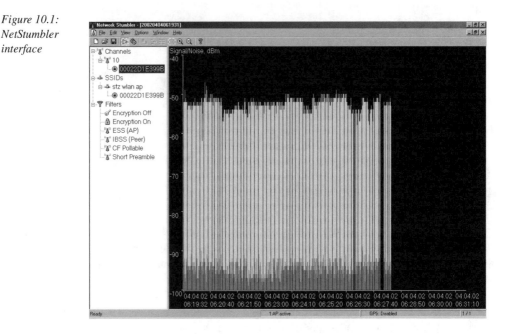

There are now a range of upgrades that ensure you can operate your systems securely. However, until now these have been manufacturer-specific, because at present there is no comprehensive standard.

## 10.1.2  Security risks

*The Wired Equivalent Privacy* (WEP) protocol in IEEE802.11 uses a static key. One of the advantages of doing so is that this kind of encryption is relatively easy to implement. However, the disadvantage is a significant security risk, because once the code has been discovered it can no longer provide any protection. There are two main methods for finding out a static key:

- Firstly, the key can be discovered by communication between humans. This is particularly critical if all the stations in a company use the same static key. However, reading the code directly from one of the mobile stations is usually not possible because it is stored on a card that uses non-volatile memory.

- Secondly an external source can use various algorithms to try to reconstruct the key.

The following scenarios should be noted here:

- The encryption of a packet starts from an *Initialisation Vector* (IV). A fixed algorithm changes this 24-bit long IV for each new packet. This means that after $2^{24}=16.7*10^6$ packets the same series of packets starts again.

- In *known plain text* attacks the attackers use known pairs of known and encrypted data to try and work out the keys that have been used. Known data can for example be derived from the known structure of IP packets or from known constants.

- In addition, the combination of stream encryption, as shown by RC4 with error recognition and correction by a linear *cyclic redundancy check* (CRC) is not secure. Not only do traceable dependencies occur between the data that is to be transferred, but also changes to the packets can remain undiscovered if the CRC data has been modified in a suitable way.

Tools such as Airsnort (http://airsnort.sourceforge.net) and Wepcrack (http://sourceforge.net/projects/wepcrack), which are free for use and can be downloaded from the Internet, are based on the weaknesses of initialisation vectors. These tools can discover what key is being used within a matter of hours. Criminals don't even need to expend much effort or have a great deal of knowledge, or ability, or use complicated tools, to do so.

There are two methods available for closing this security gap. You can change the key from time to time while still using the original WEP algorithm. Alternatively, you can use different encryption algorithms which will require more computing power to discover.

## 10.1.3   Countermeasures

Countermeasures can be implemented at various levels to close existing security gaps. They are described below.

### Complicated encryption

*Opportunities and risks*

Using more complex encryption reduces or removes the risk that eavesdropping on traffic can provide information for hackers to crack the key, because the computing power required to find out the key increases considerably. Despite this the basic risk involved in using a static, symmetrical key remains, as do the security gaps for more complicated attacks on the network.

*Standardisation*

As the shortcomings described above are also known to the IEEE, it has created a body called *Task Group i* (TGi) whose purpose is to develop and standardise a successor for WEP (see also section 4.7.3). However, it is unfortunately generally believed that this task force is not in the position to create a uniform standard. For this reason, users are currently forced to look to the different solutions provided by the various manufacturers. Below are two typical examples which are also similar to the solutions offered by other manufacturers.

*Product examples*

Various manufacturers supply products designed specifically to eliminate vulnerability to attacks via AirSnort. For example, to achieve this the company Lucent has implemented a different algorithm for generating initialisation vectors (IVs). The advantage of this solution is that you only need a driver update. As the transmitting station assigns the IV, which is transmitted in the data packet, WEPplus is backwards-compatible.

The company RSA Data security, which created the RC4 algorithm, has named its extension *Fast Packet Keying* (FPK). In it, a hashing algorithm is used to generate a unique 104-bit-long packet key for each data packet, from the constant, predefined key, the sender address, which also never changes and a packet-specific IV. Using this method, the IVs only repeat every $2^{103}=10^{31}$ packets, instead of every $2^{24}=16.7*10^6$ packets, as was formerly the case.

### EAP and 802.1X

The *Extensible Authentication Protocol* (EAP – RFC 2284) is an important basis for defining a universal and centralised security concept. Originally, EAP was developed for PPP links to provide reliable authentication for remote access users. EAP is a general protocol that provides several authentication functions. In PPP the authentication method is not selected until after the Link Control Phase (LCP) in the Authentication phase.

EAP, which was based on PPP, has now been integrated into the IEEE802.1x standard, released in 2001, which modified physical transmission over LAN networks. For this purpose EAP messages are packed into 802.1 messages (*EAP Over LAN* – EAPOL). The aim of this standard is *Port-Based Network Access Control*. Figure 10.2 shows the frame format of EAP packets packed in Ethernet:

*Figure 10.2: Frame format of EAP packets packed in Ethernet*

| Octet | Description | | Notes |
|-------|-------------|-----------|-------|
| 1 to 7 | Preamble | | |
| 8 | Start Delimiter | | |
| 9 to 14 | Destination Address | | |
| 15 to 20 | Source Address | | |
| 21 to 22 | Length / Type | | Port Access Entity (PAE) Ethernet Type |
| 23 | Protocol Version | 0000 0001 | as standard |
| 24 | Packet Type | 0000 0000 | EAP-Packet |
| | | 0000 0001 | EAPOL-Start |
| | | 0000 0010 | EAPOL-Logoff |
| | | 0000 0011 | EAPOL-Key |
| | | 0000 0100 | EAPOL-Encapsulation-ASF-Alert |
| 25-26 | Packet Body Length | | only present for EAP-Packet, EAPOL-Key, |
| 27-N | Packet Body | | EAPOL-Encapsulation-ASF-Alert |

Figure 10.3 shows the format of the EAP packets themselves:

| Octet | Description | Notes | |
|-------|-------------|-------|-------|
| 1 | Code | 1 | Request |
| | | 2 | Response |
| | | 3 | Success |
| | | 4 | Failure |
| 2 | Identifier | when combined with system port number, produces a unique identifier for authentication | |
| 3 to 4 | Length | | |
| 5 | Type | if there is a Request or Response | |
| 6 to N | Type Data | | |

*Figure 10.3: Format of EAP packets*

The EAP packet types defined up to now are listed in Figure 10.4:

| Type | Description | Source | Notes |
|------|-------------|--------|-------|
| 1 | Identity | RFC2284 | |
| 2 | Notification | RFC2284 | |
| 3 | Nak (Response only) | RFC2284 | |
| 4 | MD5-Challenge | RFC2284 | |
| 5 | One Time Password (OTP) | RFC2289 | |
| 6 | Generic Token Card | RFC2284 | |
| 7 | | | |
| 8 | | | |
| 9 | RSA Public Key Authentication | | |
| 10 | DSS Unilateral | | |
| 11 | KEA | | |
| 12 | KEA-Validate | | |
| 13 | EAP-TLS | RFC2716 | |
| 14 | Defender-Token (AXENT) | | |
| 15 | Windows 2000 EAP | | |
| 16 | Arcot Systems EAP | | |
| 17 | EAP-Cisco Wireless | | |
| 18 | Nokia IP Smart Card Authentication | | |
| 19 | SRP-SHA11 Part1 | | |
| 20 | SRP-SHA11 Part2 | | |

*Figure 10.4: Currently-defined EAP packet types*

The idea behind IEEE802.1x is that one physical connection is assigned to two logical *ports*. The physical connection always passes the received packets to the free port (*Uncontrolled Port*). However, the *Controlled Port* can only be accessed after authentication. One way is over the free port, as shown in Figure 10.5, on the next page.

*Figure 10.5:*
*Accessing the*
*Controlled Port*
*in IEEE802.1x*

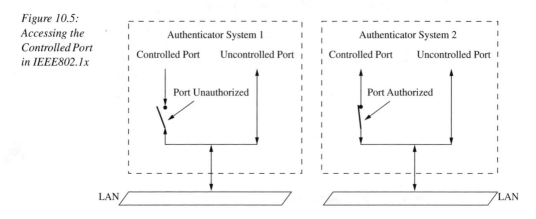

Usually a RADIUS (*Remote Access Dial-Up User Service* – RFC 2138) server acts as an authentication server. The RADIUS protocol was also designed to provide authentication for users who use a dialup connection to log on to a network. You will find a description of this protocol in RFCs 2138 and 2139, and RFCs 2865 to 2868. The EAP message is then transmitted as an attribute in the RADIUS protocol.

*Gaps in IEEE802.1x*

The IEEE802.1x standard is an important stage in the further development of a security concept for networks. However, it does have shortcomings: [Mishra (1)], in criticising it, mentions two major failings:

- the only provision in IEEE802.1x for client authentication is for the access point to release traffic over the controlled port after successful authentication. The client itself is immediately in authenticated mode. This opens the way for an attack by "the wrong server", the so-called *man in the middle attack* (see Figure 10.6):

*Figure 10.6:*
*"Man in the*
*middle attack"*
*on network*
*security*

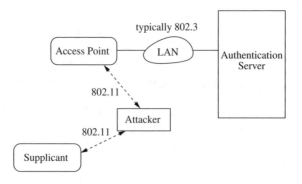

- the individual packets are no longer assigned. On this basis, a so-called *session hijacking* attack can occur, in which another station sends the successfully-authenticated client a *Disassociate* message which prompts it to close the connection. However, the access point continues to keep the controlled port open, providing the attacker with access to the network.

All the same, it is worth noting that such attacks are not very practical in the case of dialup connections, as the person dialling the telephone number of the called partner no longer needs to be authenticated. In hardwired networks, and therefore also hardwired networks that have suitable outward-facing security systems, the risk also appears relatively minor.

### Extensions

The security concept therefore still needs to be extended to cover the points mentioned above, to suit wireless networks. Firstly, *Mutual Authentication* seems an absolute prerequisite. The best-known procedures for mutual authentication include the following:

- *EAP-Transport Level Security* (EAP-TLS in RFC 2716)
- PPP *Challenge Handshake Authentication Protocol* (CHAP in RFC 1994) and Microsoft's implementation, *Microsoft Challenge Handshake Authentication Protocol version 2* (MS-CHAP v2)
- *Lightweight EAP* (LEAP) which, in Cisco systems, is also responsible for dynamically changing the WEP key.

A method is also needed to encrypt packets securely, to prevent both snooping, to find out the content of messages, and also active penetration into the network.

Cisco, for example, offers a solution on this basis for its products. It is an extended proxy server that takes on the role of a RADIUS server. At Cisco this is called an ACS (*Access Control Server*). To authenticate the WLAN terminals it uses Microsoft's *Challenge Handshake Authentication Protocol* (MS-CHAP) for the *Lightweight Extensible Authentication Protocol* (LEAP).

When LEAP is used, the logon process involves the following steps (see also Figure 10.7):

1. the IEEE802.11 client uses the uncontrolled port to log on to the access point (AP).

2. the AP blocks all requests from the client over the controlled port (such as IP requests) until that client has logged on to the network.

3. the user on the IEEE802.11 client uses their normal network logon, with their username and password, to log on to the RADIUS server, with the access point passing on their request.

4. the RADIUS server authenticates the user. To do so, it transmits an MD5 hash packet (a *Message Digest*) containing text for encryption (the *Challenge Text*) over the access point to the client. The client sends their *Response* back to the RADIUS server over the access point. In this way the client can also authenticate the RADIUS server. The client sends a text for encryption via the access point to the RADIUS server. The RADIUS server sends its *Response* via the access point to the client. The RADIUS server and client calculate the

session key which is used for WEP encryption. It is calculated using the user's password, the challenge requests and the responses from the client and server.

5. The RADIUS server sends the session key to the AP.

6. The AP encrypts its Broadcast Key with the Session Key and sends it to the client.

7. The RADIUS server and client have now mutually authenticated each other and, just like the access point, have a user-specific and session-specific key. Encrypted data transmission can therefore start.

Here it is worth noting that the transmission of the broadcast key is itself encrypted, using the session key. This key can be generated automatically, with no manual management effort, from the user name and password. In addition, keys can be changed periodically.

*Figure 10.7:*
*Logon process*
*using LEAP*

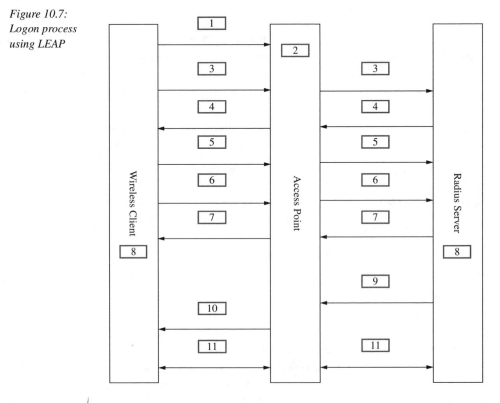

## VPNs

Using the countermeasures described above extensive security concepts, some of which already exist, can be implemented. Especially useful is the concept of the *Virtual Private Network* (VPN). VPNs are based on so-called "tunnelling" (usually layer-3 tunnelling using IPSec), but also integrate security mechanisms such as access security, authentication and encryption as integral components.

### 10.1.4  Summary

Using the security mechanisms described above, wireless LANs can be reliably secured, at least against the attack methods known today. However, this requires that the existing security measures are also really implemented. The situation is affected by the following factors:

- the available solutions are proprietary and can therefore only be implemented in a homogenous environment
- some elements of the solutions (RADIUS server, VPN, etc.) are suitable for use in larger corporate networks, but not for SoHo networks.

Which options can be recommended for use by operators of smaller and more cost-sensitive networks? They should

- definitely use any existing encryption functionality
- regularly update the static key
- download any existing new drivers with more powerful encryption, that exceed the standard, if available from the manufacturer
- assign a unique SSID to their WLAN and block SSID broadcasts
- maintain access control lists in the MAC layer
- regularly check the log files to see if they contain unknown MAC addresses, to detect any attempts at hacking into the network.

If you are willing to go to even greater lengths, you could also consider splitting the network into a WLAN segment and a "secure" segment connected via a firewall.

These measures reduce the risk of being hacked significantly, even if security gaps still remain.

In addition to that, the hacker is then forced to expend a lot of "criminal energy". That can also have its consequences, if there are legal proceedings against them at a future date.

## 10.2  Sources of interference

### 10.2.1  Introduction

When analysing active sources of interference for particular systems you should also take into account the following:

- If a receiver is in the reception range of two or more radio cells that are being operated at the same time, it needs to carry out two tasks. It must assign itself to one of the radio cells, and also suppress the signals from the other radio cells.
- Even if other communications participants are active with exactly the same protocol, without belonging to the same network, it is necessary to ensure that all stations can operate as independently as possible.

- There is a special danger in logon-free frequency ranges that other communications participants, who are using other protocols, are active on the same frequencies. Even then, support for maximum independence of operation by the different networks is still required.

- Finally interference from other systems must also be taken into account. Three sources of interference can occur:

  - other radio-based communications networks, such as GSM or radio and television transmitters which operated in another frequency range.

  - other microelectronic systems, which emit electro-magnetic radiation due to the speed of their switching procedures. These include fast-clocked microelectronic components such as microcontrollers or microprocessors, and also electronic transmitter components.

  - other devices which emit electro-magnetic radiation because of their operating characteristics. The most commonly-found devices in this group are picture tubes in televisions and computer monitors and microwave ovens.

## 10.2.2   Influence caused by identical systems

The influence caused by systems that operate using the same standard, but in separate, independent networks, involves the same issues are covered in the discussion of space repetition frequency in section 10.3.

## 10.2.3   Influence caused by competing systems

The amount of interference can be expected to rise if a number of non-co-ordinated networks using different protocols are operating in the license-free 2.4 GHz band range. The most common example of this can be seen in computers whose network functionality is implemented through IEEE802.11b-compliant radio networks and which simultaneously use Bluetooth to connect to peripheral devices. This example is also particularly critical because the distance between each antenna system may be very small. In the worst case the PC cards for IEEE802.11b and for Bluetooth may be fitted in adjacent slots in a laptop.

Very early on a number of models were created to estimate the effects of this mutual influence [Zyren (2)], [Lansford et al. (2000)]. The results of this study were based on a multitude of parameters. To summarise, it can be said that

- influence caused by the systems can be ignored if the competing devices are far enough away from each other

- influence caused by the systems becomes more noticeable if the competing devices are very close to each other

- influence caused by systems not only slows down the speed [Merritt] but can also completely disrupt functionality.

However, in a real-life situation the following points must also be taken into consideration:

- the fact that transfers can be especially badly affected if competing systems are too close together may lead to problems for many future applications. This is because laptops in particular will have both a wireless connection to the network and wireless connections to their peripheral devices. For this reason a link between the various standards in general and IEEE802.11x and Bluetooth in particular is under consideration [Shoemake].

- the fact that interference in 2.4 GHz systems currently only plays a subordinate role should not give the impression that this problem does not exist. This is because systems in open areas that operate on the basis of FDMA are particularly badly affected by this interference if dynamic frequency selection (DFS) is not supported. If these systems are located within one administrative domain, the problems can be resolved by radio planning. However, in densely populated areas serious problems can be caused by interference from neighbouring dwellings or offices in which the same frequency selection is used.

### 10.2.4   Influence caused by other radio networks

The author's own investigations about the mutual influence of radio networks implemented in other frequency ranges and wireless LAN systems in the 2.4 GHz range have shown that, as expected, no restrictions need to be observed here.

This also applies to GSM mobile telephones that operate in the 900 MHz range and to DECT networks in the 2.9 GHz range.

### 10.2.5   Influence caused by other noise pulse generators

The author's own investigations have shown that different IEEE802.11b-compliant systems and numerous standard microwave ovens have no influence on transfer quality as long as they are kept a reasonable distance apart. A slight increase in noise level could only be observed at very small distances of less than approximately 20 cm. This increase in no way caused a reduction in the data rate.

## 10.3   Selecting a spread spectrum technique

### 10.3.1   Basics

The selection of radio transfer technology also generally determines which spread spectrum technique should be used. Whereas users are only interested in the performance parameters of the overall system, some advocates of the different technologies are having a real battle about the advantages and disadvantages of the Frequency Hopping Spread Spectrum procedure (FHSS) and the Direct Sequence Spread Spectrum procedure (DSSS). There are two main reasons for this:

- Back in the days when only the IEEE802.11 standard existed, manufacturers used both these spread spectrum techniques in their own products and attempted to justify this by claiming that it improved performance.

- After the establishment of the IEEE802.11b standard, which also uses the DSSS procedure, the discussion ground to a halt because the FHSS procedure could only achieve lower data rates. However, after the FCC approved the extension of FHSS regulations as described in section 7.4.2 the manufacturers of HomeRF products moved to the FHSS camp whereas the manufacturers of IEEE802.11b-compatible products proclaim the advantages of DSSS.

## 10.3.2  Spectral efficiency

DSSS systems, as implemented in WLAN systems, have relatively low spectral efficiency. In a comparison of IEEE802.11-compliant 1 Mbps systems using the DSSS or FHSS procedure it can be observed that DSSS requires a bandwidth of 22 MHz and FHSS a bandwidth of 1 MHz per channel.

However, the space repetition frequencies selected in many practical applications of DSSS systems are higher than in FHSS systems (see also Figure 10.8). However, the greater the size of the surface area the smaller the difference.

*Figure 10.8: Physical separation of DSSS and FHSS systems*

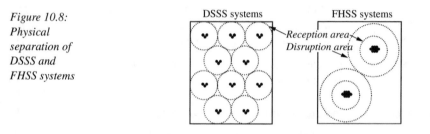

## 10.3.3  Security against eavesdropping

With regard to the frequently-stated opinion that it is harder to eavesdrop on frequency hopping procedures than direct sequence procedures, it should be noted that both systems are implemented in wireless LANs to meet FCC requirements and not to increase system security. The effect of this is that all the permitted random sequences (PRSses) are public knowledge.

- DSSS systems that comply with IEEE802.11 still only use one PN sequence. Even if IEEE802.11b is used, the possible sequences are not a secret.

- The permitted frequency sequences are also described in the standard used by slower IEEE802.11-compliant FHSS systems or HomeRF. In this way a particular frequency hopping sequence can be discovered by a comparatively simple evaluation of the time between the control signals on a few different channels [Zyren (3)].

The conclusion is that, although it is relatively easy to eavesdrop on either system, identifying FHSS signals appears to require a little more time and effort.

### 10.3.4  Sensitivity to interference

Sensitivity to the active sources of interference listed in section 10.2 is a controversial discussion topic. This issue is outlined in section 2.7. There is a range of simulation and test results that basically show that implementation at system level is the crucial factor in performance and sensitivity to interference [HomeRF Working Group].

## 10.4  Aspects of EMT interference

### 10.4.1  Basics

The biological effects of high-frequency electro-magnetic fields are a controversial subject. On the one hand, scientific investigations of physical effects have not yet been able to provide any concrete results. On the other, this subject is loaded with emotion and is politically sensitive. For this reason only a few aspects are presented: the reader is invited to make their own evaluation. The comparison with GSM mobile telephony plays an important role here. However this will only help those users who do not regard GSM as critical.

### 10.4.2  Regulations

The starting point is that all wireless LAN systems comply with the national regulations concerning transmission and field strengths. Like most other countries, the United Kingdom follows the recommendations of the *International Commission for Non-Ionizing Radiation Protection* (ICNIRP), *http:// www.icnirp.de*. In the United Kingdom, the National Radiological Protection Board is responsible for these kinds of issues. The reference value in this definition is what is known as the *Specific Absorption Rate* (SAR), which describes the absorbed output per kilogram of bodyweight. In particular this takes into account the thermal effect caused by the conversion of the absorbed radiation into heat.

Two special applications of WLAN technology can be highlighted here:

- The use of IEEE802.11-compliant systems in the area of medicine is permitted. However, once again, device-specific approval must be sought for them. Tests indicate that there are also no negative effects on the function of pacemakers [Cisco Systems, Inc.].

- Efforts are being made to implement Bluetooth systems in commercial aviation. It is generally considered that these efforts will be successful.

### 10.4.3  Comparison with GSM mobile telephony

In making a comparison with GSM mobile telephony systems the following aspects must be taken into consideration:

- *Transmitter output:* the maximum transmitter output of most wireless local networks is restricted to 100 mW. The maximum 2 W permitted in GSM mobile telephony represent a difference of more than one order of magnitude. However GSM systems use dynamic output modification so the transmitter output of GSM stations is usually well below this maximum value.

- *Positioning:* GSM mobile telephone antennae are often operated very close to the body. This is not the case for most WLAN applications.

- *Activity:* Any comparison between radio cell activity in GSM and WLAN technology depends on a wide range of parameters. In both systems it not only depends on the actual channel load but also on the activity of the entire radio cell. In particular, both systems cause a degree of base load caused by the exchange of management information.

## 10.5   WLANs and TCP/IP

Two further factor should be mentioned at this point. Both have the effect that TCP/IP-based traffic cannot be transferred over WLAN technologies at the maximum performance levels that are theoretically possible.

Firstly, TCP is a connection-oriented protocol that acknowledges the successful receipt of a packet. If this information is only transferred in one direction, as is usually the case, the receiver must then send a separate packet to the transmitter that only contains the acknowledgement. If this TCP acknowledgement is sent via a WLAN operating with the CSMA/CA algorithm the data flow shown in Figure 10.9 is the result. It is obvious that the achievable useable data rate in these cases is significantly reduced. However, no standard has yet been agreed that would permit the integration of TCP acknowledgements in CSMA/CA acknowledgements [Tourrilhes (3)].

*Figure 10.9:*
*Unidirectional*
*TCP traffic via*
*a WLAN with*
*CSMA/CA-*
*MAC.*

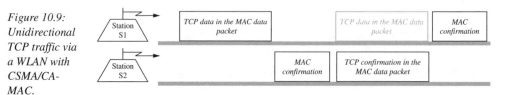

Each packet is acknowledged in the MAC and TCP layers. This results in much higher traffic levels and, in the case of continuous traffic from transmitter S1 to receiver S2, a greater risk of collisions

The second aspect is highlighted in [Schulte]: TCP/IP was developed for use in wired networks, so its time constants had to be modified to meet the requirements of wired networks. TCP therefore selects the speed of its packet transmission depending on the time from the transmission of the previous packet to the receipt of the corresponding acknowledgement (*round trip time*). As these times can be much longer in wireless networks than in wired networks, without the transmission channel actually being blocked, TCP sometimes transmits significantly fewer packets than the network could actually handle.

# 10.6 Deciding factors

## 10.6.1 Basics

Before you can select a network technology you will need to evaluate a vast array of parameters as objectively as possible. However, as subjective influencing factors can dominate, a list of deciding factors is shown below as an example. This is based on a similar procedure used when complex technical decisions must be made [Sikora (3)].

Nevertheless, when considering the criteria discussed below and shown in an overview in Figure 10.10, it must be noted that the decisions at each level should not be seen in isolation. The various criteria are actually mutually inter-dependent on each other. For example, the price of available products may influence the decision to select a particular WLAN technology.

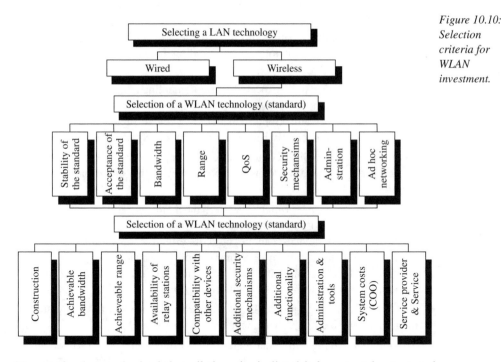

*Figure 10.10: Selection criteria for WLAN investment.*

It must also be emphasised that all the criteria listed below must be seen and evaluated against the background of the requirements of the actual application you require. For example, in an industrial application to transfer a few control signals the bandwidth required may play a less important role, whereas in a computing centre with many stations the achievable bandwidth will be a core factor in deciding for a particular system.

## 10.6.2   Selecting a LAN technology

In this context the primary decision discussed is between wired and wireless technologies. Here the basic question is how much added value (or how many savings) can be achieved for a particular application by using one or the other technology. The fundamental issues involved in answering this question have already been discussed in section 2.2.

## 10.6.3   Selecting a WLAN technology

To compare the various currently-available technologies, the following criteria (at least) should be considered:

- *Stability of the standard:* is the present version of the standard mature and stable or can additional extensions and modifications be expected which may call into question its compatibility with current equipment?
- *Acceptance of the standard:* is the standard accepted widely enough that it will be supported in future by newly-developed devices?
- *Bandwidth:* how much bandwidth is available for each particular technology? The gross and net bandwidth is the deciding factor here (see also section 2.3.2).
- *Range:* what range can be expected by implementing a particular technology? The crucial issues here are the permitted transmission output or performance density. The use of anisotropic antennae to achieve higher ranges (see also section 2.9.1) is not actually dependent on which technology is used.
- *Quality of service:* will specific services be supported? However, in this case questions need to be asked about the consistency of this support at all network levels.
- *Security mechanisms:* which security mechanisms are included in the standard? The important issue here is that only those mechanisms that are supported by all devices can be used in effective operation. As long as homogenous networks from only one manufacturer are used, the support at product level will be more than sufficient. However this scenario only applies without restrictions to point-to-point-connections such as radio bridges.
- *Administration:* which protocols are included in the standard, which allow all manufacturers to effectively manage an entire network?
- *Ad hoc networking:* to what extent is ad hoc networking functionality supported (see also section 1.3.10)?

## 10.6.4   Selecting a WLAN product

- *Construction types:* which construction types are supported? The main options for mobile PC-supported stations are USB-, PCI- or PC Card-based devices. The choice of access points is much, much greater, but usually not a critical factor.

- *Achievable bandwidths:* unlike wired systems, bandwidths largely depend on the quality of the receivers, which are not specified in the standard (see also section 1.4.7). Current test results published in the major technical periodicals provide enough information on this topic.

- *Achievable ranges:* the range over which a radio network can operate is an issue that closely relates to the question of the achievable bandwidth in each case.

- *Availability of relay stations:* in some circumstances the range of radio networks can be increased by the use of relay stations. These are mainly used for point-to-point-connections.

- *Compatibility with other devices:* even if basic device compatibility in heterogeneous environments is defined by conformity with the standard, many devices provide additional functions that are not included in the standard. Varying levels of commonality can be observed here.

- *Additional security mechanisms:* some manufacturers have added more security mechanisms to their devices, primarily when the security mechanisms included in the standard are not considered effective enough. As a result in some situations these devices can only be used in homogenous networks.

- *Additional functionality:* some devices, especially access points, provide additional network functionality such as routing or firewall functions.

- *Administration and tools:* which tools are provided for station and network management? At present it is often the case that an embedded webserver provides access via a normal web browser. However, enormous differences with regard to user-friendliness and clarity can also be observed here. Additional administration tools are usually implemented in mobile stations, which nevertheless still need a driver.

- *Price:* the cost of the entire system must be taken into consideration when determining commercial aspects. This applies not only to a comparison between wired and wireless systems, and the question of cabling and installation costs, but also to costs throughout the system's entire life cycle. System costs therefore also include administrative time and effort and any necessary upgrades or replacements.

- *Service providers and service:* last but not least, your selection, especially in large-scale investment decisions, must take into account the services that are provided with the WLAN products. In this respect, in many cases it should be ensured that the products selected come from a well-established system house.

## 10.7 Future prospects

This section ends with a glance towards future developments, even if there is the risk that the forecasts it contains are overtaken by reality and may even be contradicted:

- Wireless local networks (WLANs) will be used more and more, due to the many practical advantages they offer.
- This will drive down the cost of the active elements.
- Cheap wireless connections will lead to the development of new devices such as webpads [Bager] and new applications.
- In addition, in the medium term, developments will concentrate on just a few technologies. Until then, devices will also support several protocol standards.
- The bandwidth available for wireless transmissions will continue to increase.
- Mobile users using the Internet will increase pressure for support by the Internet protocol family [Wolisz].
- The consequence of support for ad hoc networks will be the increasing integration of intelligent functionality in networks and peripheral devices, as otherwise they will not be able to offer their own services flexibly and independently of a particular network connection.
- The increasing convergence of voice and data traffic on all network levels will lead to improved support for guaranteed quality of service. This will include WLANs. Currently, not all systems are suitable for this [Köpsel].

# A Appendix

## A.1 Maxwellian equations

The first Maxwellian equation (A.1) states that each temporally-changing electrical field generates a magnetic curl field. This is also known as the "circuital law".

$$\oint \vec{H} d\vec{s} = \int \left( \vec{j} + \frac{d\vec{D}}{dt} \right) d\vec{A}$$

(A.1)

The second Maxwellian equation (A.2) is also called the "induction law". It states that each temporally-changing magnetic field generates an electrical curl field.

$$\oint \vec{E} d\vec{s} = -\int \left( \frac{d\vec{B}}{dt} \right) d\vec{A}$$

(A.2)

When combined, electromagnetic fields are created when an electrical load is accelerated. One of Maxwell's clever predictions was that he could interpret his equations to describe the electro-magnetic waves present when no power was flowing ($\vec{j} = 0$) but both the magnetic field ($d\vec{B}/dt \neq 0$) and the electrical displacement (represented by the symbol $d$ in this formula) ($d\vec{D}/dt \neq 0$) changed.

## A.2 Physical basis of direct sequence spread spectrum process

The starting point is the narrow-band input data signal whose limit frequency $f_0$ equals the bit rate $r_{Bit}$ of the data stream. Through linking with the PN sequence of the bit rate $r_{PN}$ the signal's range of frequencies is spread in terms of frequency range, with the power of the signals $S_{PN}$ (the integrals under the two curves) remaining unchanged.

The usable power of the spread signal now results from the usable power of the original signal $S$ and the bandwidth of the PN sequence $f_0 = r_{PN}$:

$$S_{PN} = \frac{S}{r_{PN}}$$

(A.3)

Finally the signal is shifted to the high frequency range. To do so, the signal is multiplied with a high-frequency carrier signal.

At the receiver end this signal is received along with additional noise. The following operations result in the correct reception of the useful data signal:

*Figure A.1:*
*Basic structure*
*of a DSSS*
*correlator*
*(Zyren 1)*

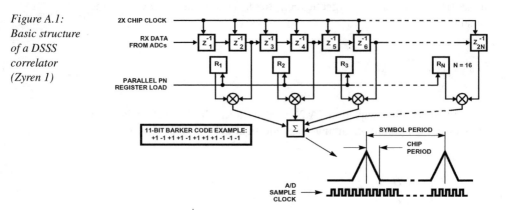

A *Matched Filter* is used to remove the overlaid *PN* sequence (Figure A.1). The basis of a matched filter of this kind is that, due to the associative law, the repeated exclusive/or operation with the same signal sequence is removed again. If *D* is the data signal and *A* is the state of the *PN* sequences, then the relationship shown in Figure A.2 applies. However, the spread functions must then have particular characteristics [Hatzold]:

• firstly, the spectra of the HF signals modulated by the spread data functions must be noise-like, which means they are evenly distributed in the frequency space. This is achieved by linking the data signals with pseudo-random functions.

• secondly, for multiple access, it is essential that the spread functions assigned to the various stations are orthogonal or at least almost orthogonal to each other. Two functions are orthogonal if their cross-correlation produces the value 0, and almost orthogonal if the cross-correlation produces a very small value when compared to the auto-correlation of one of the two functions.

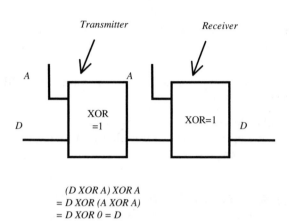

*Figure A.2*
*A DSSS*
*correlator*
*works in the*
*following way:*
*the exclusive-*
*or function is*
*used twice to*
*produce the*
*original signal*

The fundamental advantage of this process is that the narrowband interference of the high intensity $I$ is spread into a broadband noise signal of lower intensity $I_{PN}$. The resulting intensity of the interference after de-spreading is:

$$I_{PN} = \frac{I}{r_{PN}}$$

(A.4)

After low pass restriction to $f_0 = r_{Bit}$ the total noise is $I_R$, within the useable bandwidth $r_{Bit}$

$$I_r = I * r_{Bit}$$

(A.5)

This results in the following signal-to-noise ratio for the resulting signal:

$$\frac{S}{I_{PN}} = \frac{S}{I} * \frac{r_{PN}}{r_{Bit}}$$

(A.6)

This means that it is greater than the signal-to-noise ratio at the antenna input by the following spread factor: $r_{PN}/r_{Bit}$.

If a radio transmission system is to use different spread sequences to distinguish between the various channels uses, two prerequisites must be met:

- the receiver must decide which channel is intended for it. To do this, different filters receive the signal and each applies different *PN* sequences to it. The filter that has the best performance characteristics then delivers the received signal. In this filter the range of frequencies of the spread signal is then also transformed back to its original form.

- the different *PN* sequences must have a length so that it is possible to distinguish between them. As a result, the spread rate increases, which also suppresses the noise more.

A range of *PN* sequences is available, which have especially good auto correlation properties:

- Ronald H. Barker developed bit sequences that have especially good aperiodic auto correlation properties [Intersil Corp.]. The IEEE802.11 standard uses a Barker code whose length is $N = 12$.

*Table A.1:*
*Overview of*
*known Barker*
*codes*

| Code length ($N$) | Barker code |
|---|---|
| 1 | + |
| 2 | ++ or +- |
| 3 | ++- |
| 4 | +++- or ++-+ |
| 5 | +++-+ |
| 7 | +++--+- |
| 11 | +++---+--+- |
| 13 | +++++--++-+- |

- Alternatively, complex *PN* sequences are used. These are known as binary complementary codes, or sequences. The procedure used here is known as *Complimentary Code Keying* (CCK). A complementary binary code is a sequence of identical and different pairs (*like pairs:* 1 1, -1 –1, *unlike pairs* 1 –1, -1 1) with identical intervals between them. The special feature of complementary codes is that the auto correlation function of this sequence is only not equal to null if the phase shift equals null [Pearson 2000]. As a result, it is possible to implement accurate detection of the sequence.

## A.3  Directional antennae

The gain of an antenna $g(\theta)$ in the direction $\theta$ can be expressed as the relationship of the radiated power in the focussed direction to the average power to the solid angle:

$$g(\theta) = \frac{P(\theta) * 4\pi}{P_0}$$

(A.7)

Here, $P_0$ is the entire transmitted power from the antenna. The maximum output of the antenna is transmitted in the direction of the "main lobe". $g_{max}$ is the maximum antenna gain for $\theta = 0$ and indicates the gain of the antenna related to an isotropic transmitter with the same power.

For the signal power of a loss-free antenna with gain $g_s$ this produces

$$p = \frac{P_s g_s}{4\pi * r^2} = \frac{EIRP}{4\pi * r^2} \qquad \left[\frac{W}{m^2}\right] \tag{A.8}$$

The product of the power of the isotropic transmitting antenna $P_s$ and the gain from the transmitting antenna $g_s$ are known as the **Effective Isotropic Radiated Power** (EIRP). It indicates the transmitter output required for non-directed transmission with an isotropic transmitter to achieve the same usable power as would be achieved with a directional transmitter.

Table A.2 lists values achieved by typical antennae gains in real life.

| Antenna gain | Construction | |
|---|---|---|
| 21 dBi | Parabolic antenna | *Table A.2: Typical gain values achieved by different types of antenna* |
| 13.5 dBi | Yagi | |
| 5–12 dBi | Omni | |
| 6–8.5 dBi | Patch | |

The receiver now receives the signal at a strength

$$P_E = P_s g_s g_E * L = P_s g_s g_E * \left(\frac{\lambda}{4\pi * d}\right)^2 \tag{A.9}$$

Here, $g_E$ is the receiving antenna gain and $L$ is the signal attenuation in air which results from the wave length $\lambda$ and the distance $d$.

In the simplest case involving isotropic antennae, the attenuation $L_0$ results as a quotient of the received and transmitted signal power levels. In logarithmic notation we can see the difference between the two values:

$$L_0[dB] = P_E[dBm] - P_s[dBm]$$
$$= -10\log\left(\frac{P_E[mW]}{P_s[mW]}\right) = -20 * \log\left(\frac{\lambda}{4\pi * d}\right) \tag{A.10}$$

If we now include the antenna gain from the transmitting and receiving antennae, the attenuation for the entire path is:

$$L_F[dB] = -10\log\left(\frac{P_E[mW]}{P_S[mW]}\right) = -10*\log\left(\frac{P_S g_S g_E}{P_S}\left(\frac{\lambda}{4\pi*d}\right)^2\right)$$

$$= -10\log(g_S) - 10\log(g_E) - 20\log\left(\frac{\lambda}{4\pi*d}\right)$$

(A.11)

If we now set this expression to equal the equivalent attenuation and resolve it for the distance $d$, the result is:

$$L_0 = -10\log(g_S) - 10\log(g_E) - 20\log\left(\frac{\lambda}{4\pi*d}\right)$$

$$L_0 + 10\log(g_S) + 10\log(g_E) = -20\log\left(\frac{\lambda}{4\pi*d}\right)$$

$$10^{\frac{L_0+10\log(g_S)+10\log(g_E)}{20}} = \frac{\lambda}{4\pi*d}$$

$$d = \frac{\lambda}{4\pi}*10^{\frac{L_0+10\log(g_S)+10\log(g_E)}{20}}$$

$$d = \frac{1}{4\pi}*\frac{c}{f}*10^{\frac{L_0+10\log(g_S)+10\log(g_E)}{20}}$$

(A.12)

Here, $c$ is the transmission speed which can be set to equal the speed of light as a first approximation.

Equations (A.10) to (A.12) are based on identical transmission powers of 1 mW. If we now increase the transmission power of the isotropic transmitter by

$$p_{SV} = \frac{\text{diffused power }[mW]}{1mW}$$

(A.13)

then this fact must also be taken into consideration. The result is:

$$d = \frac{1}{4\pi}*\frac{c}{f}*10^{\frac{L_0+10\log(p_{SV})+10\log(g_S)+10\log(g_E)}{20}}$$

(A.14)

Let us now use the following typical values in equation (A.14): $L_0 = 90$ dBm, $g_S = g_E = 6$ dBi and $p_{SV} = 13$ dBm. If we allow a maximum margin of 10 dBm, this produces a range of 2190 m. Naturally, this requires a direct *Line of Sight* connection (LOS)) and sufficient separation from any obstructions on the ground so that the waves can propagate without difficulty, and the assumed attenuation characteristic is achieved as far as possible.

By using directional antennae of this kind the WLAN systems described can also be used for linking physically separate LANs. However, in the geographical area in which ETSI applies, and when the 2.4 GHz band is being used, the limiting factor is that the amplification gain and antenna gain must produce a signal of less than +20 dBm EIRP, with a transmitter output of 1 mW and an isotropic transmitting antenna.

# B Bibliography

## B.1 English-language publications

This section lists the publications, which are referenced in this book, that are in English. Section B.2 contains details of publications and other information sources in German, that are referenced in this book.

Aramaki, T., "Difference between BRAN HiperLAN2 and MMAC HiSWANa", *HIPERLAN/2 Global Forum, 21st February 2001*

Arbaugh, W.A., Shankar, N, Wan, Y.C.J., "Your 802.11 Wireless Network has No Clothes", *University of Maryland, 30th March 2001*

Abramsson, N., "The ALOHA system – another alternative for computer communications", *AFIPS Conference Proceedings, Vol. 37, FJCC 1970, pages 695–702*

Bluetooth standard (*http://www.bluetooth.com/developer/specification/Bluetooth_11_Specifications_Book.pdf* and *http://www.bluetooth.com/developer/specification/Bluetooth_11_Profiles_Book.pdf*)

(1) Borisov, N., Goldberg, I., Wagner, D., "Security of the WEP Algorithm", available from: *http://www.isaac.cs.berkeley.edu/isaac/mobicom.pdf*

(2) Borisov, N., Goldberg, I., Wagner, D., "Intercepting Mobile Communications: The Insecurity of 802.11", Seventh Annual International Conference on Mobile Computing And Networking", 16th to 21st July 2001, available from: *http://www.isaac.cs.berkeley.edu/isaac/mobicom.pdf*

Burkley, C. J., "Coding and Modulation Techniques for RF LANs", in: Santamaría, A., López-Hernández, F.J., "Wireless LAN Systems", published by: *Artech House, Boston, London 1994, ISBN 0-89006-609-4, pages 197–224.*

Cisco Systems, Inc. "White Paper: Cisco Systems Spread Spectrum Radios and RF Safety", available from: *http://www.cisco.com/warp/cc/pd/witc/ao340ap/prodlit/rfhr_wi.htm*

Compaq, "Compaq Wireless LAN", available from: *http://www.compaq.de/wireless/technologie/lan/index.htm*

Constantinou, P., "Properties of RF Channels", in: Santamaría, A., López-Hernández, F.J., "Wireless LAN Systems", published by: *Artech House, Boston, London 1994, ISBN 0-89006-609-4, pages 129–163*

FCC Part 15 (*http://www.fcc.gov/Bureaus/Engineering_Technology/Orders/2000/fcc00312.txt*)

Fluhrer, S., Mantin, I., Shamir, A,. "Weaknesses in the Key Scheduling Algorithm of RC4", Preliminary Draft, 25.7.01, available from: *http://www.crypto.com/papers/others/rc4_ksaproc.ps*

Frodigh, M., Johannson, P., Larsson, P., "Wireless ad hoc networking – The art of networking without a network", Ericsson Review No. 4, 2000, pages 248–263, available from: *http://www.ericsson.com/review/2000_04/files/2000046.pdf*

(1) Frost & Sullivan, "World Wireless LAN Markets", #5781-74, 1999. Extracts available from: *http://www.proxim.com/wireless/research/frost_sul.pdf*

(2) Frost & Sullivan, "Bluetooth boosts demand for chips", press release, January 2001 (*www.frost.com*)

Haartsen, J.C., Zürbes, S., "Bluetooth voice and data performance in 802.11 DS WLAN environment", *Ericsson, 31st May 1999*

Halford, K., Webster, M., "Multipath Measurement in Wireless LANs", Intersil Application Note AN9895, May 2000, available from: *http://www.intersil.com/data/an/an9/an9895/an9895.pdf*

HomeRF Working Group, "Interference Immunity of 2.4 GHz Wireless LANs", available from: *http://www.homerf.org/data/tech/hrf_interference_immun_wp.pdf*

Intersil Corp., "A Condensed Review of Spread Spectrum Techniques for ISM Band Systems", Intersil Application Note AN9820, 1st May 2000, available from: *http://www.intersil.com/data/an/an9/an9820/an9820.pdf*

Johnson, M., "HiperLAN/2 – The Broadband Radio Transmission Technology Operating in the 5GHz Frequency Band", HiperLAN/2 Global Forum, 1999, available from: *http://www.hiperlan2.com/web/pdf/whitepaper.pdf*

Khun-Jush, J., Malmgren, G., Schramm, P., Torsner, J., "HiperLAN type 2 for broadband wireless communication", in Ericsson Review No. 2, 2000, pages 180–119, available from: *http://www.ericsson.com/review/2000_02/files/2000026.pdf*

Köpsel, A., Wolisz, A., "Voice transmission in an IEEE 802.11 WLAN based access network", in Proceedings WoWMoM 2001, Rome, Italy, July 2001, available from: *http://www-tkn.ee.tu-berlin.de/publications/koepsel/wowmom02.pdf*

Lansford, J,. Nevo, R., Monello, B., "Wi-Fi™ (802.11b) and Bluetooth Simultaneous Operation: Characterizing the Problem", available from: *http://www.mobilian.com/documents/WPSG_2.pdf*

Merritt, R., "Conflicts between Bluetooth and wireless LANs called minor", EE Times, 21st February 2001, available from: *http://www.planetanalog.com/story/OEG20010220S0040*

Mettala, R., "Bluetooth Protocol Architecture", *Version 2.0, 25 August 1999, Bluetooth Special Interest Group, Doc.No. 2.C.120/2.0*

Mishra, A., Arbaugh, W.A., "An Initial Security Analysis of the IEEE802.1X Standard", *University of Maryland, 6th February 2002*

Pearson, B., "Complementary Code Keying Made Simple", Intersil Application Note AN9850, 1st May 2000, available from: *http://www.intersil.com/data/an/ an9/an9850/an9850.pdf*

Proxim Inc., "Migration of Wireless LAN Technologies in the Enterprise", White Paper 2001, available from: *http://www.proxim.com/products/harmony/white papers/migration.pdf*

Shoemake, M.B., "WiFi (IEEE802.11b) and Bluetooth – Coexistence Issues and Solutions for the 2.4 GHz ISM Band", Texas Instruments, White Paper, February 2001, Version 2.1, available after logon at: *http://www.ti.com/sc/wire lessnetworking*

(1) Solectek Corp., "Tech Talk – DSSS vs. FHSS, The Skinny", available from: *http://www.solectek.com/tech-center/index.html*

(2) Solectek Corp., "Tech Talk – What's in Your Microwave", available from: *http://www.solectek.com/tech-center/index.html*

Spurgeon, C. E., "Ethernet – The Definitive Guide", published by: *O'Reilly, 2000, ISBN 1-56592-660-9*

Texas Instruments, "TI's Optical Wireless Solutions™ Technology – Moving Data at Light Speed", available from: *http://www.ti.com/sc/docs/products/msp/telecom/ alp/ows_brochure.pdf*

(1) Tourrilhes, C.R.J., "Linux Wireless LAN Howto", available from: *http:// www.hpl.hp.com/personal/Jean_Tourrilhes/Linux/Linux.Wireless.pdf*

(2) Tourrilhes, C.R.J., "A Medium Access Protocol for Wireless LANs which supports Isochronous and Asynchronous Traffic", available from: *http:// www.hpl.hp.com/personal/Jean_Tourrilhes/Papers/pimrc98.html*

(3) Tourrilhes, C.R.J., "PiggyData: Reducing CSMA/CA collisions for multimedia and TCP connections", *proceedings of VTC `99. HP external report HPL-1999-58*

Wireless Ethernet Compatibility Alliance, "802.11b Wired Equivalent Privacy (WEP) Security, 19 February 2001, available from: *http://www.wi-fi.com/pdf/Wi-FiWEPSecurity.pdf*

Wolisz, A. "Wireless Internet Architectures: Selected Issues. Mobile Networks, pages 1–16, 2000, (invited paper) in: J. Wozniak, J. Konorski, editors "Personal Wireless Communications" Kluwer Academic Publishers, Boston/Dordrecht/ London, available from: *http://www-tkn.ee.tu-berlin.de/publications/wolisz/ pwc_wolisz_cor.pdf*

(1) Zyren, J., Petrick, A., "Brief Tutorial on IEEE 802.11 Wireless LANs", Intersil Application Note 9829, February 1999, available from: *http:// www.intersil.com/data/an/an9/an9829/an9829.pdf*

(2) Zyren, J., "Reliability of IEEE802.11 Hi Rate DSSS WLANs in a High Density Bluetooth Environment", *Intersil, 8th June 1999*

(3) Zyren, J., Godfrey, T., Eaton, D., "Does Frequency Hopping Enhance Security", 19th April 2001, available from: *http://www.wi-fi.com/pdf/20010419_frequencyHopping.pdf*

## B.2    German-language publications

The publications below, which are referenced in this book, are in German.

Ahlers, E., Živadinovic, D., "Datenkuriere – 12 WLAN-Systeme für schnelle Funknetzwerke", published in: *c't magasine, issue 22, 2000*

Andren, C., "Der Teufel steckt im Detail – Wireless LANs in Gebäuden einrichten", published in: *Design&Elektronik magazine, issue 6, 2001, pages 86–89*

Bager, J., "Die Web-Pads kommen! – Drahtlos surfen ohne PC", published in: *c't magasine, 2001, issue 16, pages 120–122*

(1) Bläsner, W., "Blaubeere an Blauzahn, bitte kommen! – Mit System zum Markterfolg bei Bluetooth", published in: *Design&Elektronik magasine, issue 6, 2001, pages 72–76*

(2) Bläsner, W., "DECT: Revolution durch LIF-Technik", published in: *Elektronik magasine, issue 7, 2000, pages 74–76*

Gieselmann, H., "Sendung mit der Maus – Neun Funkmäuse und fünf kabellose Tastaturen im Test", published in: *c't magasine 2001, issue 11, pages 154–165*

Hascher, W., "Höchstintegration bei DECT-Bausteinen", published in: *Elektronik magasine, issue 9, 1999, pages 52–53*

Hatzold, P., "Spread Spectrum und CDMA – Was dahinter steckt", published in: *Elektronik magasine, issue 21, 2000, pages 56–64*

Heeg, M., "Vernetzung ohne Schnur", published in: *Elektronik magasine, issue 10, 1999, pages 68–74*

Kaiser, W., script for lectures called "Geschichte der Technik im Industriezeitalter I" and "Geschichte der Technik im Industriezeitalter II", given at *RWTH Aachen,Germany, 1999*

Kammeyer, K.D., "Nachrichtenübertragung", published by: *B.G. Teubner, Stuttgart, 1992, ISBN 3-519-06142-2*

Klinkenberg-Haass, F., "Bluetooth quo vadis?", article published 13th June 2001, available from: *http://www.tecchannel.de/hardware/724/index.html*

Lebherz, M., Wiesbeck, W., "Beurteilung des Reflexions- und Schirmungs-verhaltens von Baustoffen", published in: *Bauphysik 12, 1990, issue 3, pages 85–92.*

Lüke, H.-D., "Signalübertragung – Grundlagen der digitalen und analogen Nach richtenübertragungssysteme", 4th edition, published by: *Springer Verlag, 1990*

Mazzara, A., "10 vor 10 vom Donnerstag 24.5.2001". This is a television programme, broadcast by the Swiss television station "DRS", available from: *http://real.sri.ch/ramgen/sfdrs/10vor10/2001/10vor10_24052001.rm?start= 0:01:37.905&end=0:10:15.612*

Nett, E,. Mock, M., Gergeleit, M., "Das drahtlose Ethernet – Der 802.11 Standard: Grundlagen und Anwendung", published by: *Addison-Wesley - Data com Akademie, 2001, ISBN 3-8272-1741-X*

Schulte, G., "Fallstricke – TCP/IP im drahtlosen Netz", published in: *c't magasine, issue 6, 1999, pages 232–233*

Sietmann, R., "Quo vadis, Mobilfunk? – Nebenbuhler und Nachfolger von UMTS", published in: *c't magasine, issue 5, 2001, pages 94–101*

(1) Sikora, A., "802.11: Standard für drahtlose Netze (standard for wireless networks", article published 12.04.2001, available from: *http:// www.tecchannel.de/hardware/680/index.html*

(2) Sikora, A., "DECT – Die Alternative zu Bluetooth (DECT - the alternative to Bluetooth", article published 21.8.01, available from: *http://www.tecchannel.de/ hardware/511/index.html*

(3) Sikora, A., "Programmierbare Logikbauelemente – Architekturen und Anwendungen", published by: *Hanser Verlag, 2001*

(4) Sikora, A., "Embedded Applikationen im Internet, Teil 1: Übersicht über Vor- und Nachteile von vernetzten Anwendungen", published in: *Elektronik magasine, issue 22, 2000, pages 90–102*, "Teil 2: Implementierungen",. published in: *Elektronik magasine, volume 23, 2000, pages 164–169*

(5) Sikora, A., "CMOS goes to Gigahertz – Embedded Technologien werden schneller", published in: *Markt & Technik magasine , issue 22, 2000, pages 39– 40*

Vinayak, V., O'Keese, W., Lam, C., "Ein Chip für die Übertragung im 2,4 GHz- Band – Eine gemeinsame Lösung für Bluetooth und Funk-Daten-netze", part 1, *Elektronik magasine, issue 7, 2001, pages 70–77*

Walke, B., "Mobilfunknetze und ihre Protokolle", published by: *Teubner 2000, Volume 1, ISBN 3-519-16430-2* and *Volume 2, ISBN 3-519-16431-0*

Wuschek, M., "Elektrosmog: Gefahren durch Mobilfunk?", article published 12th January 2001, available from: *http://www.tecchannel.de/hardware/628/index.html*

# Abbreviations

| | |
|---|---|
| ACK | Acknowledgement |
| ACL | Asynchronous Connectionless Link |
| ACS | Access Control Server |
| ADSL | Asymmetric Digital Subscriber Line |
| AID | Association Identifier |
| AM_ADDR | Active Member Address |
| AP | Access Point |
| APC | Access Point Controller |
| API | Application Program Interface |
| AR_ADDR | Access Request Member Address |
| ARIB | Association of Radio Industries and Broadcasting |
| BD_ADDR | Bluetooth Device Address |
| BER | Bit Error Rate |
| BPSK | Binary Phase Shift Keying |
| BRAN | Broadband Radio Access Network |
| BSIG | Bluetooth Special Interest Group |
| BSS | Basic Service Set |
| BT | Bandwidth Time |
| CA | Collision Avoidance |
| CC | Central Controller |
| CCA | Clear Channel Assignment |
| CCK | Complimentary Code Keying |
| CD | Collision Detection |
| CDMA | Code Division Multiple Access |
| CEM | Client Encryption Manager |
| CEPT | European Conference for Post and Telecommunications |
| CFP | Contention Free Period |
| CL | Convergence Layer |
| CM | Central Mode |
| CODEC | Coder/Decoder |
| CP | Central Point |
| CRC | Cyclic Redundancy Check |

| CS | Carrier Sense |
| CSMA/CD | Carrier Sense Multiple Access |
| CTS | Clear-To-Send |
| CVSD | Continuous Variable Delta Modulation |
| CW | Contention Window |
| DAB | Digital Audio Broadcasting |
| DAC | Device Access Code |
| DCF | Distributed Coordination Function |
| DCS | Dynamic Channel Selection |
| DECT | Digital European Cordless Telecommunication |
| DECT-MMC | DECT Multimedia Consortium |
| DFS | Dynamic Frequency Selection |
| DH | Data–High Rate |
| DIAC | Dedicated Inquiry Access Code |
| DLA | Direct Link Access |
| DLC | Data Link Control |
| DM | Data–Medium Rate |
| DM | Direct Mode |
| DMAP | DECT Multimedia Access Profile |
| DPRS | DECT Packet Radio Service |
| DQPSK | Differential Quadrature Shift Keying |
| DRM | Dynamic Resource Management |
| DS | Distribution System |
| DSSS | Direct-Sequence Spread Spectrum |
| DV | Data Voice Combined |
| EC | Error Control |
| ERIP | Effective Isotropically Radiated Power |
| ESS | Extended Service Set |
| ETSI | European Telecommunications Standard Institute |
| FCC | Federal Communications Commission |
| FDMA | Frequency Division Multiple Access |
| FEC | Forward Error Correction |
| FFT | Fast Fourier Transform |
| FHSS | Frequency Hopping Spread Spectrum |
| FP | Fixed Part |
| FSK | Frequency Shift Keying |
| FTP | File Transfer Protocol |

| | |
|---|---|
| GAP | Generic Access Profile |
| GFSK | Gaussian Phase Shift Keying |
| GIAC | General Inquiry Access Code |
| GIP | DECT/GSM Interworking Profile |
| GP | Processing Gain |
| GPRS | General Packet Radio Service |
| GSM | General-System-for-Mobile-Communication |
| H2GF | HiperLAN/2 Global Forum |
| HCI | Host Controller Interface |
| HIPERLAN/2 | High Performance Radio Local-Area Network Type 2 |
| HiSWANa | High Speed Wireless Access Network |
| HomeRF | Home Radio Frequency |
| HomeRF WG | Home Radio Frequency Working Group |
| HR | High Rate |
| HV | High Quality Voice |
| IAPP | Inter Access Point Protocol |
| IBSS | Independent Basic Service Set |
| ICNIRP | International Commission for Non-Ionizing Radiation Protection |
| ICV | Integrity Check Value |
| IEEE | Institute of Electrical and Electronic Engineers |
| IFFT | Inverse FFT |
| IFS | Interframe Space |
| IN | Intelligent networks |
| IP | Intellectual Property |
| IP | Internet Protocol |
| IrDA | Infrared Data Association |
| ISI | Intersymbol Interference |
| ISM | Industrial, Scientific, Medical |
| ISO | International Standards Organisation |
| ISP | Internet Service Provider |
| ITU | International Telecommunication Union |
| IV | Initialisation Vector |
| KSA | Key Scheduling Algorithm |
| L2CAP | Logical Link Control and Adaptation Protocol |
| LAN | Local Area Network |
| LBT | Listen Before Talk |
| LLC | Logical Link Control |

| LMP | Link Manager Protocol |
| LOS | Line of Sight |
| MA | Multiple Access |
| MAC | Medium Access Control |
| MBOK | M-ary Bi-Orthogonal Keying |
| MC | Multi Carrier |
| MLS | Multi-Level Signalling |
| MMAC | Multimedia Mobile Access Communications |
| MMAP | Multimedia Access Profile |
| MSK | Minimum Shift Keying |
| MT | Mobile Terminal |
| NAV | Net Allocation Vector |
| OBEX | Object-Exchange |
| OFDM | Orthogonal Frequency Division Multiplex |
| OSI | Open Systems Interconnection |
| P2P | Point-to-Point |
| PAN | Personal Area Network |
| PC | Point Coordinator |
| PCF | Point Coordination Function |
| PCM | Pulse Code Modulation |
| PDA | Personal Digital Assistant |
| PDU | Protocol Data Unit |
| PHS | Personal Handy-Phone System |
| PLCP | Physical Layer Convergence Protocol |
| PM_ADDR | Parked Member Address |
| PMD | Physical Medium Dependant |
| PMP | Point-to-Multipoint |
| PP | Portable Part |
| PPM | Pulse Position Modulation |
| PPP | Point-to-Point-Protocol |
| PRS | Pseudo Random Sequence |
| PSK | Phase Shift Keying |
| 64QAM | Quadrature Amplitude Modulation |
| QoS | Quality of Service |
| QPSK | Quadrature Phase Shift Keying |
| RADIUS | Remote Authentication Dial-In User Service |
| RAP | Radio Local Loop Access Profile |

| RegTP | **Reg**ulator for **T**elecommunications and **P**ost |
|---|---|
| RLC | **R**adio **L**ink **C**ontrol |
| RSSI | **R**eceived **S**ignal **S**trength **I**ndication |
| RTS | **R**equest-**T**o-**S**end |
| SA | **S**tandards **A**ssociation |
| SAP | **S**ervice **A**ccess **P**oint |
| SAR | **S**egmentation **A**nd **R**eassembly |
| SCO | **S**ynchronous **C**onnection-**O**riented Link |
| SDMA | **S**pace **D**ivision **M**ultiple **A**ccess |
| SDP | **S**ervice **D**iscovery **P**rotocol |
| SN | **S**ervice **N**egotiation |
| SNMP | **S**imple **N**etwork **M**anagement **P**rotocol |
| SNR | **S**ignal-**T**o-**N**oise **R**atio |
| SOHO | **S**mall **O**ffice **H**ome **O**ffice |
| SSID, ESSID | **E**lectronic **S**ystem **ID** |
| SST | **S**pread **S**pectrum **T**echniques |
| SWAP-CA | **S**hared **W**ireless **A**ccess **P**rotocol – **C**ordless **A**ccess |
| TBTT | **T**arget **B**eacon **T**ransmission **T**ime |
| TCP | **T**ransport **C**ontrol **P**rotocol |
| TCS binary | **T**elephony **C**ontrol **S**ervice-**Binary** |
| TDD | **T**ime **D**ivision **D**uplex |
| TDMA | **T**ime **D**ivision **M**ultiple **A**ccess |
| TPC | **T**ransmit **P**ower **C**ontrol |
| TSF | **T**iming **S**ynchronisation **F**unction |
| UDP | **U**ser **D**atagramm **P**rotocol |
| UNII | **U**nlicensed **N**ational **I**nformation **I**nfrastructure |
| USB | **U**niversal **S**erial **B**us |
| WECA | **W**ireless **E**thernet **C**ompatibility **A**lliance |
| WEP | **W**ired **E**quivalent **P**rivacy |
| Wi-Fi | **Wi**reless **Fi**delity |
| WLAN | **W**ireless **L**ocal **A**rea **N**etworks |
| WLANA | **W**ireless **LAN** **A**ssociation |
| WLIF | **W**ireless **LAN** **I**nteroperability **F**orum |
| WLL | **W**ireless **L**ocal **L**oop |

# Index